D1359769

PURCHASED FROM
MULTNOMAH COUNTY LIBRARY
TITLE WAVE BOOKSTORE

The Entrepreneurial Engineer:
Starting Your Own High-Tech Company

For a complete listing of the *Artech House Technology Management and Professional Development Library*, turn to the back of this book.

The Entrepreneurial Engineer:
Starting Your Own High-Tech Company

R. Wayne Fields

Artech House
Boston • London

Library of Congress Cataloging-in-Publication Data
Fields, R. W. (R. Wayne)
  The entrepreneurial engineer : starting your own high-tech
company / R. Wayne Fields.
     p. cm. — (Artech House technology management and professional
development library)
  Includes bibliographical references and index.
  ISBN 1-58053-029-X (alk. paper)
  1. Engineering firms—Management. I. Title. II. Series.
TA190.F55 1999   99-36698
620'.0068—dc21   CIP

British Library Cataloguing in Publication Data
Fields, R. Wayne
  The entrepreneurial engineer : starting your own high-tech
  company. — (Artech House technology management and
  professional development library)
  1. High technology industries 2. New business enterprises
  1. Title
  338.4'76

   ISBN 1-58053-029-X

Cover design by Lynda Fishbourne

© 1999 ARTECH HOUSE, INC.
685 Canton Street
Norwood, MA 02062

All rights reserved. Printed and bound in the United States of America. No part of
this book may be reproduced or utilized in any form or by any means, electronic or
mechanical, including photocopying, recording, or by any information storage and re-
trieval system, without permission in writing from the publisher.

   All terms mentioned in this book that are known to be trademarks or service marks
have been appropriately capitalized. Artech House cannot attest to the accuracy of this
information. Use of a term in this book should not be regarded as affecting the validity of any trademark or service mark.

International Standard Book Number: 1-58053-029-X
Cataloging-In-Publication: 99-36698

10 9 8 7 6 5 4 3 2 1

To my loving wife, Lorraine,
and wonderful daughter, Nicole.
I'm very proud to be part of us.

Contents

Preface

I spent the better part of 10 years as president and co-owner of a respected business development company in Lake Oswego, a suburb of Portland, Oregon. Basically, we were venture capitalists for the little guys, although we operated on a business plan different from that of traditional venture capitalists. In fact, Venture Solutions, Ltd. (VSL) is still in place, but less active because my partner and I have moved on to other activities leveraged by the original business. (VSL's toll-free number is 1-800-346-2408.)

VSL is unique in many ways, one being the focus on what we call the privately funded tier of business development complexity. VSL's interest is in the substantial midrange entrepreneurial sector, addressing projects that are larger than the average retail shop but smaller than the typical project of appeal to venture capitalists. In other words, we deal with the situation that confronts most practicing engineers or other technical people who have a product idea: how to turn their notion into a successful high-growth, product-oriented company.

Our approach is widely applicable because we raise few prerequisites. We show how to carry forth a primordial product idea, acquiring all the necessary money, talent, and resources to build a successful company, and we even show how to develop a founding idea from scratch if one is not yet at hand. If you have a sincere desire to form a business and make it succeed, the business development recipe presented in this book will show you how. Along the way, this introductory treatment points to virtually all the important issues that must be addressed and the general approach to their resolution.

The Entrepreneurial Engineer focuses on a defined series of temporal stages to ascend from a raw idea to a profitable business. Each business development stage is fundamental to the overall story, because project makeup, dynamics, and challenges change drastically with advancement from a primordial notion, through product and business development, and on to major-company status. While some companies come to emphasize one internal facet or another, such as marketing, product design, or production, an important mindset for understanding the book's message is to realize that every facet of a company is of prime concern and must be dutifully mastered. Success is not a product of chance or luck, even if fortuitous things happen along the way. Experience shows that meticulous attention to every detail of the recipe presented in *The Entrepreneurial Engineer* is far and away the best means to enhance the probability of success. Nothing can guarantee a win, but if you continuously look, act, and think success and steadfastly posture the project in position for success, chances are you eventually will achieve a win.

Entrepreneurship is about taking risks to attain rewards. This book spends little time touting potential rewards, such as personal wealth, satisfaction in building a company, technical marketing and business accomplishment, creating jobs, and the respect of others. It is assumed that you already appreciate the possible rewards and that your real questions center on how to build a product idea into a successful high-growth business. *The Entrepreneurial Engineer* is designed to introduce and properly relate virtually all the main challenges that must be confronted, giving you the perspective to know the correct issues to address and questions to ask at every stage of business development. In addition, the Suggested Reading section at the end of the book provides innumerable sources of additional information regarding diverse relevant subjects.

R. Wayne Fields
Santa Barbara, California
August 1999

Acknowledgments

I would like to thank the in-house and outside reviewers at Artech House, especially Barbara Lovenvirth and Tina Kolb, for untold help in transforming technical thought into an accessible, coordinated, and properly marketed whole. The efficient and thorough treatment of all aspects of manuscript and book preparation by these true professionals was outstanding. Furthermore, I must assume all responsibility for any remaining errors or mistakes of composition, fact, interpretation, or clarity.

Part I
Overview of Business Development

The Entrepreneurial Engineer is about a journey. It begins with a novel idea, a flash of insight, and ends when that idea runs its course. The journey can be very short or can last a lifetime. You are the pilot, and the journey is your trajectory through the world of business development, an environment where riches can be earned but that is harsh and unforgiving for those who do not heed its demanding conventions and rules. This book details unique business development principles and practices applicable to product-oriented, high-growth, technically based startups typical of engineers and other technocrats with commercially viable ideas.

Part I is an overview of the total business development process and provides specific orientation to several key issues central to later discourse. The treatment goes beyond fact and implementation to teach a mental set best suited to designing, building, and carrying out the targeted business development effort. This critical mental frame is essential to properly understanding the process on which you embark and the way to confront the countless significant events that undoubtedly will arise. History shows that the cavalier will fail, and that careful accommodation and constant readjustment of a formal, projectwide strategic plan of action within a properly structured and managed company are the secrets of success.

Chapter 1 begins by defining the business development process and the special niche of that process. Privately funded projects are the focus, a specific subset of business development most relevant to the target business form. Also included is a whirlwind tour of business development, and background to show how the current approach has been designed from real-world experience to really work. Chapter 2 then introduces seven paramount success factors used throughout the book; a project built around those factors alone is powerfully postured for success. Finally, Chapter 3 closes Part I with an intermediate-level coverage of six defined business development stages, augmentation of background material, and development of a language and mental frame as foundation for detailed discussion of each business development stage later in the book.

Part I is preparatory for the detailed processes, methods, and rationale that form the remaining three parts, the book's goal being to provide a complete introductory recipe for building a winning project. Factual development is considered essential, but the real object is building the added benefit of insight, carrying repeated discussion of important ideas through new and progressively more complex situations. The approach here derives from a decade and a half of experience in entrepreneurial projects like yours. Experience shows that lists of rules and facts are not enough; it is necessary to create a functional web of interrelated facts and relations before real understanding begins.

1

A View of the Overall Process

1.1 Ideas and Their Management

Let's begin with an idea for a product, a concept relating to a service, or perhaps a product/service mix. It is a great feeling to create a novel idea that also portends real commercial potential.

While the feeling may be exciting, the thought of bringing the project to fruition may be daunting. The average engineer or other technical person may be highly skilled in one or more areas and may even be a talented technical manager, but project success depends on far more. *The Entrepreneurial Engineer* presents a recipe that shows exactly how any technical person can carry out the complex process of business development.

It is good news when you may have created a truly unique idea, with the potential for personal and business success. This book shows you how to analyze your project for viability and to adjust its structure to better ensure success. If you do not have a beginning idea, *The Entrepreneurial Engineer* shows how to develop new ones on demand, totally from scratch.

You also will learn that few projects actually prosper. If that seems harsh, especially stated so early in the text, there is a good reason. This book forms a stark reality check with regard to countless classes of project pitfalls. The hard lesson is that someone must control the project from conception to completion, attention to every detail being essential. By following the recipe given here, any person with a basic technical background can design and implement any project. Project scope, the extent of personal knowledge and experience, cash, or immediate access to human or technical resources are not central to success. But you must follow the recipe with dutiful—almost obsessive—attention to all issues, both large and small. Business development is extremely unforgiving.

Most projects fail, not from bad ideas, but because most enterprises do not follow time-tested methods of development, structure, and management. The book's cardinal theme is to show how to understand and carry out the long list of actions and methods on which virtually every successful project depends.

Common sense does not often reign in the entrepreneurial trenches. However, a surprising number of universal success factors mark virtually all winning projects. Beyond that, many more project-specific factors exert substantial added impact. This book carefully explores a wealth of points, both universal and particular, combining descriptions with pertinent examples. The goal is to provide broad and deep practical understanding of the real-world matters to be faced at every project stage.

The fool depends on luck, but the recipe given in these pages defies luck. The main thrust is to show you how to plan, to prepare, and to cover every base. That can be accomplished by anyone who meticulously adapts an individual project to the book's detailed methods.

It is critical that you, the founder, are the one person who constantly ensures that all project bases are covered, even if, at some point, others are placed wholly or partly in charge. Leave nothing to chance; addressing most, but not all, issues will not do.

This book is strictly directed to descriptions and examples of product-based business development programs. Given the primary audience of engineers and other technically oriented people, most fertile ideas will be product related, so a book product focus provides the maximum audience benefit. Those with service ideas still can profit substantially from the book. Most product-oriented discussion and examples have a direct service analog or at least can be extrapolated to a service business design by direct, common-sense analogy.

Projects usually do not fail because the idea was poor. That possibility is thoroughly assessed and corrected along the way and is not an issue in well-run endeavors. Projects almost always fail due to insufficient attention to *all universal* and *all relevant particular* details. On the basis of 15 years of employment, consulting, and partnership in entrepreneurial enterprises, involving hundreds of observed projects, the author has learned that practically any problem can be traced most fundamentally to human origin.

1.2 Privately Funded Projects

The ultimate spectrum of successful businesses forms a huge range of size and complexity. The smallest category, termed here *proprietary shops* (e.g., bakeries, auto parts stores, nonanchor mall outlets, metal and wood fabricators, and single-owner shops), is not the subject of this book. The other business extreme is the very large project appropriate for formal venture capital, prominent contemporary examples residing in the fields of genetic engineering and electronic commerce. Such *venture capital projects* also are not in the book's domain. The third and final business category, *privately funded projects*, covers the broad arena between the low-end (usually small and simple) proprietary shops and the high-end (commonly large and complex) venture capital projects. Examples of privately funded projects include Starbucks Coffee® and Oil Can Henry's®.

Although seemingly limited, the realm of privately funded projects is not all that confined. True, there are many forms of proprietary shops, and some extend upward in complexity to rival privately funded projects. The privately funded project's lower border is fuzzy, and its upper border is blurred as well, due to analogous overlap with smaller venture capital projects. But privately funded projects are distinguished not only by size but by their nature as well.

This book has little to say about the proprietary shop or the venture capital business complexity planes. The privately funded complexity tier has its own rules for success, and that is the focus here. Superficial similarities exist among all three levels of business complexity, but this text circumvents surface resemblance in its minute and critical analyses of successful business development. The goal is to equip you with a compelling understanding of and tools for the specific privately funded project niche. Our job is to arm you with comprehensive theory and practice, thoroughly road tested in real-world privately funded projects, placing you in the enviable position of controlling your own destiny. The alternative, typically endured by the unprepared, is being ruled by an out-of-control project that seems to have a mind of its own.

Privately funded projects are unique, as shown by the following list of major distinctions.

- *Products.* Products, not services, are the most common focus of privately funded projects. Venture capital projects are similar in their focus, but services are the main thrust of proprietary shops.
- *Planning.* Planning is critical but is seldom given much attention in privately funded projects. It is virtually ignored by proprietary shops but is mandatory and elaborate for venture capital projects.
- *Management.* In privately funded projects, management is critical, but the full importance is rarely understood. Most proprietary shops virtually ignore management, but venture capital endeavors give it high importance.
- *Funding.* Funding is paramount for privately funded projects, but acquisition methods are poorly understood. Most proprietary shops lack adequate professional knowledge or guidance, while funding is high

priority for venture capital projects, which bring major sophistication to bear.

- *Marketing.* Marketing is crucial but often slighted in privately funded projects. Proprietary shops develop methods primarily through guessing or experimentation, whereas marketing in venture capital projects typically enjoys intense focus and highly professional (and expensive) methods.

- *Location.* A storefront is rarely required for a privately funded project, so location is usually keyed to materials or customers. Proprietary shops usually depend on storefront and/or relative immediacy of market, while venture capital enterprises span the full spectrum of possibilities.

This list appears again, in Section 2.2, which describes seven success factors, six of which are those in the preceding list. This correspondence of lists exemplifies a bookwide effort to standardize concepts. The entire business development process can seem overwhelming at first. Actually, the process is straightforward, although multifaceted. Distinguishing features of *The Entrepreneurial Engineer* include consistent terminology and maximal simplification of the business development process without loss of content and coherence.

The preceding list highlights substantial differences between privately funded businesses versus both the (typically) smaller proprietary shop and the (usually) larger venture capital enterprise. Historically, "how-to" books and articles have addressed the proprietary shop tier, while many major books, implicitly or explicitly, have addressed the venture capital tier. To date, what is defined here as the privately funded realm has been largely ignored, a serious omission. There is a large universe of creative engineers and allied professionals who constantly produce ideas with exciting commercial potential. These people need only real professional guidance in the vastly larger business development realm. Privately funded projects are complex, dynamic entities. Success dictates bringing to bear carefully integrated expertise from a broad spectrum of disciplines.

The latter point may seem intimidating, but this book does not expect you to carry out the multitude of progressive, diverse project processes alone. With hardly an exception, the days of one-person entrepreneurial efforts are over. The wise inventor recruits a team with broad-based expertise and skills, and this book shows you how to do just that effectively and efficiently.

Of course, you may be another Bill Gates, someone who can conjure what appears to be magic every day. But if not, this book is designed for you; it thoroughly addresses overall essentials but also treats a wealth of fine practical details derived from hundreds of real-world projects. These collective efforts span an enormous range of sizes, markets, players, and many other dimensions that an academic survey would likely shortchange. *The Entrepreneurial Engineer* prepares you for the snarling tangle of the real world.

Many of you have been in a foreign country where you did not speak the language. Unaccompanied by someone acting as interpreter, you probably felt quite isolated at times. That situation resembles a skilled engineer wishing to benefit from commercializing a solid product idea. Fortunately, with proper preparation, you can exert substantial control and direction on entrepreneurship, regardless of whether you have sufficient business development knowledge or access to various human, technical, marketing, business, and financial resources. All you must do is follow the guidelines presented here, being sure to implement each step relevant to your particular project.

1.3 What Is a Good Idea?

Not every idea is a winner. Idea viability must be thoroughly assessed along a number of dimensions. Here are five tests against which every idea should be rigorously challenged.

- *Duplication.* Has the idea been developed before in a fashion that basically conforms to the subject idea?
- *Similarity.* Even if the idea has not been precisely duplicated previously, has the spirit of the concept been captured elsewhere in some meaningful way?
- *Feasibility.* Given the present state of knowledge, is the notion forbidden by natural law or otherwise physically beyond theoretical embodiment?
- *Producibility.* Even if the product concept is feasible, is it beyond current capabilities of practical embodiment?
- *Marketability.* If the product notion is both feasible and producible, does it meet practical criteria for profitable marketing and sale?
- *Legality.* Is the product legal?

A given idea may pass or fail one or more of those idea tests. A pass means the idea should continue through the business development process outlined in the chapters to come, where it will be retested many times and from multiple perspectives. Failing one or more tests, however, is not necessarily a death blow; ideas often can be altered to pass a test posture.

Note that the tests of an idea are rather broad. Every listed item is really an obstacle class rather than a single hurdle. To provide further insight, the following subsections detail the identified obstacle classes.

1.3.1 Duplication

An idea essentially equivalent to an existing commercial product may be subject to rejection. It is reasonable to conclude that if everything about a concept is basically the same as an existing product, a development effort is unwarranted and unwise.

For our purposes, let us label any existing commercialized product as "Brand X." Assume that Brand X already has an effective design, has met all regulatory hurdles (if any), is in production, and enjoys effective marketing that has generated much product name recognition. As Snapple® discovered relative to the soft drink industry, those entering a successful market face severe challenges. Snapple® had an innovative product line, employed broad professional marketing, and used tons of money in their effort to buy into a large, profitable existing market. Yet the original Snapple® effort, actually far larger than even the traditional venture capital project, was a failure. The Snapple® product line was then sold, and time will tell if the new owner can establish a successful niche in a mature market where the competition certainly does not take such encroachment lightly.

On the other hand, aspects of an existing Brand-X business may provide some leverage for potential "wannabes." For decades, U.S. automakers primarily offered large cars with poor economy and dubious quality. When the Japanese entered the American auto market, they defined it differently and apparently offered many superior alternatives. The small, efficient, reliable Japanese cars managed to capture and hold a substantial U.S. market share.

These illustrations convey a point relevant throughout the book. A global view is essential, a given issue requiring extended assessment relative to the whole project and even the entire company and marketplace environment.

For simplicity, you first analyze the subject question, here product duplication, based purely on the limited issue itself. Ultimately, you must reevaluate all decisions and plans in the global context of the entire business considered over extended time.

The referenced global analysis is one of many compelling reasons to conduct the detailed planning and business plan activities described in Chapters 5 and 6. Such planning provides a comprehensive independent analysis of all project aspects, including the original idea. Every project aspect is evaluated in great detail and from many perspectives. Further, the planning process forces careful integration of all project aspects at every descriptive level. In a real sense, planning constitutes a reality tuneup, not only on each project facet independently but for a proper melding of all facets into a coherent whole.

1.3.2 Similarity

Like idea duplication, an analysis of idea similarity may or may not compel rejection. Products can be viewed as existing in a multidimensional space. For instance, common dimensions would include function, cost, target customer, geographical market, market channels, manufacturing approach, distribution scheme, and so forth. The list could be extended without obvious limit, simply by reducing the distinction between product parameters or by using a longer parameter list.

Idea similarity evaluation depends on choosing useful analysis dimensions (Table 1.1). The process compares the idea relative to competitive units (e.g., Brands X, Y, and Z) to assess distinctiveness. First, list every dimension considered relevant to the analysis. That may sound difficult and arbitrary now, but this book is designed to recursively highlight the requisite data so the basic concepts become second nature. While actual lists might differ from person to person, the substance of this book would lead most people to generate lists covering all the key bases.

Idea duplication and idea similarity are so alike that the idea similarity multidimensional analysis also can be applied to idea duplication (and the other idea tests). In fact, multidimensional analysis can be useful in many diverse contexts.

Interpreting multidimensional analysis can be aided by prioritizing the dimensions list (see Table 1.1). Place the most important dimensions for your particular product and its environment first on the list, positioning remaining dimensions below in order of importance. Next, assign each item a weighting multiple reflecting relative importance. Multiples ranging from 1 to 2 or 1 to 3 are usually best, but narrower or broader ranges may be more appropriate on occasion. Assign your product a score between 1 and 10, 10 being best, for every listed dimension. Repeat the process for each competitive brand being analyzed. Finally, for every listed dimension, multiply each product's raw score by that dimension's weighting factor to calculate the weighted product scores. Summing each product's weighted score column generates a total grade indicative of relative overall product strength.

It is impossible to provide a universally applicable list order, the variance of different products being far too great. However, this book recursively presents key concepts, instilling the insight you need to personally create, from scratch, sound lists as needed.

Table 1.1
Prioritized Multi-Dimensional Analysis

Prioritized Dimension List	Company Product	Brand X	Brand Y	Brand Z	Dimension Weighting	Weighted Company	Weighted Brand X	Weighted Brand Y	Weighted Brand Z
Price (sales literature)	10	8	7	6	3.0	30.0	24.0	21.0	18.0
User ease (marketing research)	10	10	7	4	2.8	28.0	28.0	19.6	11.2
Quality (technical evaluation)	10	8	6	4	2.8	28.0	22.4	16.8	11.2
Product direct mfg cost (technical estimate)	10	8	8	7	2.7	27.0	21.6	21.6	18.9
Customer support (marketing research)	10	9	7	7	2.6	26.0	23.4	18.2	18.2
User appeal (marketing research)	9	8	5	3	2.5	22.5	20.0	12.5	7.5
User documentation (lab evaluation)	10	9	10	8	2.5	25.0	22.5	25.0	20.0
Warranty (sales literature)	10	10	10	10	2.4	24.0	24.0	24.0	24.0
Size (lab measurement)	9	8	7	9	2.3	20.7	18.4	16.1	20.7
Weight (lab measurement)	8	10	5	8	2.3	18.4	23.0	11.5	18.4
User serviceability (marketing research)	9	7	8	4	2.0	18.0	14.0	16.0	8.0
Technical support (marketing research)	10	7	7	5	2.0	20.0	14.0	14.0	10.0
Return policy (marketing research)	10	9	10	10	1.8	18.0	16.2	18.0	18.0
Product design cost (technical estimate)	8	6	9	8	1.8	14.4	10.8	16.2	14.4
Packaging (lab evaluation)	10	9	9	9	1.5	15.0	13.5	13.5	13.5
Technical documentation (lab evaluation)	9	7	6	2	1.5	13.5	10.5	9.0	3.0
Tooling & mfg setup cost (technical estimate)	6	4	10	6	1.0	6.0	4.0	10.0	6.0
Market share—factory direct sales (mkg res)	0	10	8	6	0.0	0.0	0.0	0.0	0.0
Market share—distributor channel (mkg res)	0	10	0	5	0.0	0.0	0.0	0.0	0.0

Table 1.1 *(continued)*

Weighted Score (Relative Product Strength)	354.5	310.3	283.0	241.0
Present analysis assumes an expensive electromechanical company product entering a market of three entrenched competitors. Company product is imaged and positioned as superior in features and quality but competitively priced. The score sets for certain dimensions do not include a best (10) score, due to existing industry weaknesses amenable to future improvement. Zero dimension weightings effectively remove the item from current analysis (irrelevant for this analysis).				

1.3.3 Feasibility

Our current state of knowledge and technology can obstruct tangible idea embodiment in two ways. First, current understanding of physical law may not permit material representation of, say, a concept that includes an antigravity component. Second, while permitted in principle, a concept may be beyond current capabilities to specify a working product, such as designing a robot with human intelligence. Infeasible product ideas must be identified and rejected early, to avoid wasting valuable human, technical, and financial resources.

Initial infeasibility, however, need not spell doom. Look further for ways to appropriately modify the idea, pulling it into the realm of physical law or practicability, a notion applicable to all other idea-screening tests under discussion. If your original idea includes an antigravity element, find another force generator or eliminate the need for force all together. If human-level artificial intelligence existed in the original product idea, reconfigure the system to involve, say, a computerized expert system and a human operator working together. By using successive reanalyses, you often can recast the concept so that the impossible becomes realistic.

1.3.4 Producibility

Ideas may be amenable to imagination and design but infeasible to manufacture. Device concepts and designs exist today, such as motors and gears, that are truly molecular in dimensions. But these visions exceed current fabrication and assembly techniques, existing instead in the futuristic domain termed *nanotechnology*. Nanotechnology resides several orders of magnitude below microtechnology, a realm entered only recently and certainly not conquered. For now, products employing some degree of nanotechnology must be deemed nonproducible. Within a few decades, though, we will be able to essentially "grow" many nanotechnology products one molecule at a time.

1.3.5 Marketability

For completeness, we must address ideas that are both feasible and producible but unmarketable. Consider a product that has no meaningful market that can be reasonably targeted with acceptable cost. A hypothetical illustration might be a device that must be priced very low to sell but that targets a small and disparate population segment not easily distinguished for mailings (e.g., left-handed, color-blind consultants), and mailings may be too expensive anyway. Assume the only other choice is broad-based nonspecific advertising (e.g., radio, TV), a method impractical due to low selling price and concomitant lack of large promotional resources. We have a desired, producible, and design-feasible product, but it is basically unmarketable.

Alternatively, the product may be something that applies to an accessible population segment but is still unmarketable due to one or more unmanageable aspects. Examples of those aspects are high price (e.g., a super pencil sharpener that must retail for $1,000) and obligatory negative features (e.g., a home insect destructor that creates foul odors).

1.3.6 Legality

This book's theory and methods are designed to mesh legally into society. The prescribed approach carefully blends the limited product view with that of the broader company environment. Total company environment extends far beyond direct product considerations to include such factors as funding, competition, economic climate, regulatory issues, social context, and, implicitly, legality.

Although legality may seem morally obvious, it is also essential for other reasons. The theory and methods described in this book will not work with illegal products or unlawful business practices. For one, almost all potential investors would no longer be interested.

Another critical consideration under the global legal umbrella is regulatory affairs. Many products are not subject to any regulations, while others are. Because miscues can kill a project, early identification, assessment, and resolution of all applicable product regulatory issues are paramount for two reasons. First, such action avoids later disasters of project delay or termination due to regulatory agency intervention. Second, regulatory constraints can affect product design, production, storage, distribution, or marketing, more subtle aspects that catch many projects off guard. Regulatory compliance must be system-inherent from the very start, not cobbled on later. Quality must be addressed precisely the same way.

1.4 The Business Development Process

The extended activity of turning an idea into a business is termed *business development*, a process that relies mainly on business and science, with perhaps a touch of art.

Many would-be entrepreneurs believe they can simply wing it, reacting to issues as time unfolds. Nothing could be further from the truth. This book demonstrates that a casual approach is likely to lead to disappointment. In addition to highlighting virtually all classes of pitfalls and potholes that await the unprepared, we show how to design and carry out a program that avoids most trouble and effectively deals with the unexpected. This is not idealistic academic musing; the theory and methods described here come straight from real-world experience. The topics and problems addressed are exactly those that aspiring entrepreneurs face in the universe of privately funded projects.

Figure 1.1 is a composite of the traditional domestic business development process as described by others. Precise business development segregation is somewhat arbitrary, since process staging could be increased or decreased simply by adjusting stage breakpoints. The overall business development process is really a continuum, so any staging is arbitrary but illuminating for discussion. In addition, particular stage labels match widely employed terms and concepts, are likely familiar to the practicing engineer or allied technical person, and should correlate well with their professional experience.

This discussion and Figure 1.1 are merely the tip of the iceberg. The focus of *The Entrepreneurial Engineer* is a blanket dissection of the entire business development process, including how the process should work; why, how, and where it can go wrong; and countless illustrations of real-world problems and solutions. Finally, the treatment carefully emphasizes unique nuances of the privately funded project, our main focus.

For perspective, Chapter 2 continues presentation of broad introductory material and provides a comprehensive introduction to seven universal business development success factors. To complete Part I, Chapter 3 presents a detailed overview of the entire business development process. Part II, Professional

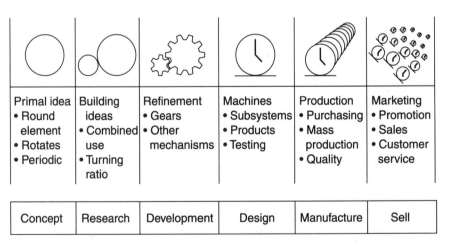

Primal idea	Building ideas	Refinement	Machines	Production	Marketing
• Round element	• Combined use	• Gears	• Subsystems	• Purchasing	• Promotion
• Rotates	• Turning ratio	• Other mechanisms	• Products	• Mass production	• Sales
• Periodic			• Testing	• Quality	• Customer service

Concept	Research	Development	Design	Manufacture	Sell

Figure 1.1 Generic schematic of the traditional business development process. Top row: abstract sketches of hypothetical clock development. Bottom row: elements of the relevant staged product development process.

Planning and Funding (Chapters 4 through 7), delivers a penetrating examination of detailed concept creation and management, planning, the business plan, and funding, all in a realistic project and business context. The latter four topics, absolute linchpins of winning projects, are subsequently applied to business development's key central stages in Part III, Product Design and Launch. Finally, Part IV, Building Long-Term Value, wraps up business development and how to instill control and growing value on the maturing company.

Following the book's main body are two appendixes integral to the global message. Appendix A presents a detailed discussion of the business plan, an essential ingredient that in regularly updated form affects, drives, and supports every business development stage. Appendix B presents a "war story," a real-world example of a privately funded project involving the promulgated business development process.

1.5 The "Founder-As-God" Trap

From experience, one overriding issue compels early mention and repeated emphasis. Even if everything else is solid, this one issue is often a project killer. It occurs in perhaps 20% of projects generally and seems even more prevalent among highly technical founders. The basic problem is inductive reasoning, a trap central to much seductive but bad thinking by people operating beyond their expertise. A skilled engineer may have false delusions of, say, marketing, administrative, manufacturing, quality, or funding prowess as well. You, the founder, must constantly ensure that the project is free of this deadly defect, an issue hard to assess objectively because of personal involvement. Some inventors think, quite incorrectly, that they are the only ones who can bring their idea forth. Inventor arrogance can be extreme, and its presence typically dooms a project.

Two types of self-tests can help founders and other key players judge project strength and correct identified weaknesses. The two tests are *project attributes* and *personal attributes* and actually apply to any project, whether a founder-as-god condition exists or not.

Here are several prominent project attributes where trouble is particularly easy to spot:

- *Project control.* "Project gods" tend to insist on far more project control than their experience or capabilities warrant.
- *The team.* Project gods usually pick other core team members who are relatively weak, probably to reduce opposition.

- *Funding.* The ability to secure funding is severely hurt by project gods, but the project gods themselves usually deny any problem exists and the remaining team members will not raise the issue.
- *Ownership.* Project gods greatly overrate their value, demanding far too much ownership share.
- *Planning and the business plan.* A formal business plan, updated regularly and based on solid strategic and tactical planning, is a must. Project gods deny plan value, and any "planning" is purely mental and more reactive than proactive.
- *The business development process.* The general approach of project gods is winging the entire development process; the disastrous result is a project trajectory that often resembles a random walk more than a controlled, midcourse-adjusted trajectory.
- *Project value.* Often overlooked, a logically unfolding project creates added value through documentation, control, and consistency; project gods devastate value.

The preceding checklist identifies key whole-project attributes that are essentially mandatory for project success. It is instructive to parallel the project attribute analysis with the following list of personal attributes that are essential in a successful founder.

- *Delegation.* Be able to delegate substantial and, in rare cases, total responsibility, especially at advanced operational stages.
- *Planning.* Do the detailed planning as background, write and regularly update a written business plan, and always use the plan framework for discussion and decisions.
- *Sharing.* Be willing to part with an ownership majority to ensure a strong team and sound funding. All else follows from there.
- *Control.* Carefully structure the formal business, team makeup, and responsibilities. Form a board of directors that is strong, functional, and motivated and that exudes those traits to outsiders.
- *Leadership.* Build and retain a truly strong project management team and then respect them. If you are not the best leader, place one or more team members above you and appropriately get out of their way.
- *Self-positioning.* Wisdom includes knowing what you do not know. From that foundation, make sure the project team covers every base with top talent and that you and all others are correctly positioned based on actual ongoing functional role.

- *Team dynamics.* Extending the self-positioning attribute, ensure, at every project advancement stage, that management makeup meets advancing complexity and properly anticipates future need.

- *Professional help.* Retain capable and respected professional legal and accounting support as soon as feasible. Financial records must be accurate, complete, and accessible.

- *Quality.* Build the company and its products on a sound quality assurance (QA) and quality control (QC) base designed into the business and offerings at inception.

Those two checklists, project attributes and personal attributes, represent two main dimensions of any company. One can view a company as the legal person to which it collectively equates or as the collective of people who run it.

In assessing the project, it is imperative for you as founder to conduct a truly objective self-examination. To accomplish that, first carry out an honest self-appraisal and a realistic project assessment. Second, seek independent assessments, which can be extremely constructive. Self- or outside analysis of either or both checklists easily can be ambushed by founder manipulation. Such manipulation is self-destructive but prevalent. Self-deception makes a weak project even weaker, because many identifiable problems are not addressed.

The structure of project assessment is crucial. It is obvious that either the project attributes test or the personal attributes test alone would convey distinct project value. However, the best project examination uses both checklists. Adapt every issue from the combined lists to create a master register, which we will call the *company profile*, tailored to your project. That provides a penetrating assessment and a robust foundation for project design and implementation.

The suggested evaluation is not absolute and has not been predictively validated in any way. It is simply an informal means to roughly gauge personal and project status and need. The combined roster of select project attributes and personal attributes is key. Properly designed and conducted, the company profile provides an invaluable standard for project design and implementation, because company profile elements greatly correlate with real-world project success.

The project profile is meant to highlight project weakness and guide deficiency resolution. The first step is assessing project stance (profile) by cataloging human and other resources. Then allocate human and nonhuman resources to best accommodate the current project plan. That may reveal plan modification opportunities to better leverage present project resources. Using the adjusted project plan, allocate existing resources and identify missing human and nonhuman resources. Finally, design and execute a plan to secure all missing resources. *The Entrepreneurial Engineer* shows how all that can be done, even if at first you, your resources, and money are limited.

1.6 Deciding to Commit

The discussion to this point contains an unstated but paramount implication, that the founder must make a full commitment to the project. It must be emphasized that the level of commitment required is likely far beyond anything you have done before, with few if any exceptions. You not only will be married to the project, but it could grow to become the most palpable expression of your professional identity.

Treating the project with complete professionalism and respect and following the methods outlined in this book, you can grab and retain maximal project control throughout its tenure. It is the foolish that buck professional savvy and rely on luck.

Most privately funded projects require sizable commitments by several key players in addition to the project founder. While usually less than that of the founder, the requisite commitment by project team members can be substantial even though they typically have much less to gain. As the book progresses, that potential problem is resolved in fairness to all.

1.7 The Road Ahead

Section 1.4 described the layout of chapters and subjects to come. The goal here is to reinforce a formal, professional mindset that has been a major chapter objective. Adopting a mental frame consistent with the book's recipe is essential for responding appropriately to the substantial challenges of business development. In contrast, many past works, mostly targeting proprietary shop or venture capital project types, are to varying degrees academic in tone. This book incorporates relevant academics, too, but specifically in a privately funded, product-oriented context supported by real-world examples.

Realize that the blunt approach here is not meant to induce fear, but to command full attention. While the lessons are demanding, they are also manageable. The chief problem is the vulnerability of startup companies, which lack depth to survive mistakes and so must do things right the first time. Be proactive by planning and managing for success. Depending on luck and reactive responses is not professional management, but amateur incompetence.

Introductions aside, the time has come for business development process specifics. Our approach is multifaceted, treating the broad structural and operational problems and solutions in business development from both an academic and practical perspective.

The business development recipe described in this book is founded on proactive, not reactive, project structure, design, and management. That means carefully planning ahead, avoiding a constant reeling from unanticipated but mostly predictable events. Outlined is a complete, professional approach to create and build any privately funded project, emphasizing acquisition and retention of project control from inception, through development, and over subsequent commercialization stages. Each business development stage is detailed relative to action and rationale. Finally, we show how to assemble a project team that virtually cannot be denied, capable of fulfilling carefully conceived plans, acquiring necessary funding, and otherwise interacting properly with the outside world. In other words, *The Entrepreneurial Engineer* not only postures you to carry out preplanned action but to achieve your plan in the face of inevitable surprises.

2

Success Factors

2.1 Common Denominators of Failure

Most entrepreneurial projects fail twice. First, they fail to exploit tremendous conventional wisdom about business development, and then the business fails.

Picture yourself in a place so foreign to past experience that you are quite dismayed and disoriented. To the uninitiated technical person, business development is similar, full of potential but dangerous. Unfortunately, both traveler and budding entrepreneur see much potential but do not perceive the many dangers. Observations by the author suggest that both overconfidence and ignorance play major roles.

People often fall back on common-sense reckoning rather than sound judgment and careful preparation to guide plans and actions. Expertise and common sense may serve well within one's general profession and private world, but it often is inappropriate, even counterproductive when used outside familiar territory. Business development is a realm where normal common sense works poorly and can be disastrous. This author has watched countless inexperienced people commit major errors of judgment and action, oblivious to danger signs that would be obvious to experienced business developers or successful entrepreneurs.

The prevalence of such problems suggests a search for common denominators of failure. The discussion here frames three large issues, based on the author's years of experience dealing with diverse entrepreneurial projects within the privately funded tier of project complexity.

- *Short on understanding.* Most people begin a project with deficient knowledge and erroneous preconceptions about the overall business development process. This defect relates fundamentally not only to

21

lack of information but to people being primarily dependent on insufficient preparation to properly identify and resolve project structure, management, operation, and funding.

- *Short on resources.* Projects are overwhelmingly found short on essential resources, such as money, talent, and equipment. But the real problem—again—is fundamentally a people one, involving miscalculation of all elements required to sustain the effort long-term.

- *Short on reality.* People often believe, idealistically, that the original idea will sell itself. The product is perceived as so great that success should be obvious to investors. At heart, this is another people problem: The project founder subscribes to a common-sense but incorrect model of what fuels a successful project.

These common project failure modes are clearly people dependent. Ideas do not fail; people fail to properly develop them. This message is a major theme of *The Entrepreneurial Engineer.*

It has been said that the only thing wrong with experience is that there is only one way to get it, an adage that conveys considerable truth. For you, it means acquiring the needed expertise in all pertinent project areas. We are not suggesting you spend years of personal study and research. We also are not suggesting that you hire the needed skills and knowledge, although that is one simple (but usually impractical) approach. What we recommend is for you to build a project team that, along with your personal strengths, covers all knowledge and skills crucial to your unique project. That may sound difficult, but it is more a matter of applying persistence and discipline in implementing the recipe given in this book.

Let's turn from a look at failure and analyze what breeds success.

2.2 Seven Factors for Success

Every project is so different that studying a few reveals little global likeness. But applying keen hindsight scrutiny to a large spectrum of endeavors, we have culled and tested a number of key ingredients of success.

By "tested," we mean taking a concept derived from one project and applying it to many others in a type of thought experiment. We derive a tentative template of behavior from one setting and then imagine the template laid over other settings (projects) to see if common patterns are apparent. This is, at most, an inexact science, but due attention and practice can lead to useful results.

This section focuses on seven success factors extracted from real-world experience. These success factors represent a major theme in this book, addressed over and over from various points of view. This list represents the core elements around which a project must be conceived, built, and guided:

- *Perspective.* Nothing can substitute for the deep and broad insight that progressively accrues from constant vigilance by savvy leaders.
- *Planning.* Planning provides a stern but flexible frame of reference against which all decisions and actions, large and small, must be judged.
- *Management.* Governing authority provides the essential glue that employs skill, experience, and perspective to turn promise and hope into tangible value and cash flow.
- *Funding.* Securing the right amount of money at each of many critical stages in the birth and growth of a business is so paramount that little proceeds without it.
- *Product.* The most basic elements that serve to set project identity and worth are the design of the products that form the nucleus of future cash flow.
- *Marketing.* Besides investment, the only money that comes into the ideal startup company is the sole result of the critical function of marketing.
- *Location.* Location is a wild card whose impact depends on the relation between the product mix and the raw materials and markets for each product.

These factors do not represent a unique order of relative importance, since projects differ in the impact of individual success factors. The nuances of any one project can certainly alter the rank of criticality and therefore due focus. As usual, the project founder must consider and apply major and minor points presented here in light of the subject project. The goal is to arm you with a broad array of facts, methods, and instincts often counter to common sense but absolutely central to wise project direction at every development stage.

Figure 2.1 depicts the business development process in terms of the seven success factors. The figure is laid out as a pyramid, to emphasize the strength and stability of properly conceived and implemented projects. Most projects, however, might be more aptly depicted as upside-down pyramids, vulnerable to myriad disturbances.

With that introduction, the rest of this section examines each success factor in detail.

Figure 2.1 Seven success factors of the business development process.

2.2.1 Perspective

The other six factors are specialties in their own right, each having supported whole careers and huge businesses. But the first success factor, perspective, is different. Perspective means the deep and broad grasp on the instantaneous state of the whole project, taken against the context of near- and far-term plans and prospects. It is thus a multidimensional view of everything past, present, and future melded together into a vague but coherent whole.

You have the ability, in a very real way, to exert control on the nature of your journey through business development. Sure, the discipline is hard, and it takes talent, determination, and persistence to succeed. We will not downplay the required effort and endurance you must exhibit to bring a new project forth. But given proper preparation, the journey can be accomplished, and you can reap some intangible personal benefits, along with the monetary rewards, that just cannot be achieved otherwise.

Perspective is a secret ingredient to success. That is true, in one sense, because success can be measured in other ways besides wealth. But in addition, success is bred by a project discipline and insight that goes beyond money and that also transcends well-laid and -executed plans. To optimize success, you must be in a position to stick to the discipline of project planning but also be ready to occasionally bend with the wind. We will dissect planning in great detail, being one of the seven success factors. It is extremely important to follow the planning that has been carefully laid out, as you will see later. But here, we are discussing an even higher plane of wisdom and insight that can come only from long-term project immersion with a decidedly global view. If this book's recipes are followed, you will develop a sixth sense that flags opportunities or dangers at the slightest hint, a manifestation we call *perspective*.

If you carefully adhere to the processes and procedures developed throughout this book, you surely will develop automatically a measure of true perspective. You probably will not even notice it is happening, but occasionally you will take note of signs that you are gaining perspective. One tipoff occurs when you have dutifully kept actual activities on plan, but new insight arises that suggests key changes could improve plan results. Or you may begin to recognize unified patterns of secondary or tertiary consequences during planning that suddenly go far beyond previous experience. These are distinctive signs that perspective's wisdom is becoming your companion. Revel in personal perspective and cultivate it in others as an invaluable ally.

2.2.2 Planning

Of the six specialized success factors, planning is an imperative that will be repeatedly highlighted as critical. It deserves and in fact requires such emphasis, because planning is so misunderstood by the technical and nontechnical alike. Most observed entrepreneurs have failed to realize the need for detailed, recursive planning; many in fact feel it is wasteful. We sincerely hope this book summarily puts that destructive idea to rest.

Some people appreciate planning as a proactive means of supporting goal achievement. However, while intentions are good, many of that persuasion just never seem to get around to planning. Others deem planning as lower priority from the start; they prefer to deal reactively with things as they arise. For both groups, perhaps to differing degrees, planning appears to interfere with more important issues of daily toil. A third group, those who truly appreciate the imperative of planning, represent a small minority.

Probably the most valuable time you will spend is in careful project planning. Chapter 5 presents many impressive reasons for detailed planning and for cross-checking plans already made. For now, we offer one big reason: If you do not do the planning, you will not get any money.

2.2.3 Management

After planning, the most important of the six specialized success factors is management. Management is important for both internal and external reasons. Within the project, management is crucial to devise, stage, implement, lead, and follow up on all activities, large and small. Without management, coordination, command, and control (C^3) just do not happen, and the project almost certainly will go astray. Management is essential externally too, the biggest element being money. You will not get any money without a powerful management (project) team.

In this book, *project team* and *management team* mean the same thing. Although the label makes each term sound like an entity, what you really should picture is a process. While the management team is made up of people, and people are bona fide entities, management's real aim is to instill coordination, command, and control, a process. Management is a lot more than a collection of figureheads residing atop company departments or functions. Rather, view management as a precisely integrated set of interweaved and interleaved functions that design and bring order to all project activities. In addition, effective management involves leadership, that intangible quality to point the way, explain its importance, define achievement means, and show how this will benefit the company and its people. Successful management is the glue that employs knowledge, skill, and perspective to unify the project and drive it to productive fulfillment.

Optimal management makeup may change from time to time because the project passes through many, many development stages. When a product idea is conceived, management may consist of the founder alone. The founder may immediately involve others, at least by discussing the product, or may keep the idea under wraps, cautiously developing it in secret for fear of losing the opportunity. In any case, to convert the idea into a true project, the primal thought must be successively elaborated into project development plans. As the original product notion matures into a project, the structure of management usually changes as well. More broadly, to maintain effective direction in the face of drastic environmental changes, projects typically must undergo many changes of management style and perhaps makeup as the project matures in purpose, direction, and complexity.

Management is key to obtaining money, and investment is a young project's life-blood. Thus, ignoring internal benefits, management is paramount because of external forces alone. Consider the ideal investor, one who presumes that a management team smart enough and strong enough can bring almost any project to fruition, even faced with wide variances of actual product merit. We have seen teams of savvy managers spin off from successful companies (e.g., Tektronics, Intel, Apple Computer) and obtain sizable investment capital from private investors before they have even identified a first product for pursuit. That scenario is rare and should not be the basis of the average project strategy. Nevertheless, it underlines a profound truth, that management strength and savvy carry weight far beyond the common-sense notion of filling privileged boxes in an organizational chart. The makeup of management can easily make or break a project.

2.2.4 Funding

Now we turn to funding. We cannot overstate the essential fact that if you have no money, you have no project.

Many budding entrepreneurs make a common error. They see product value as self-evident. In their minds, it is obvious that if their product were marketed today, it virtually would sell itself. Excessive enthusiasm transforms into presumptions that the founder's vision permeates everyone else too, including other project team members, vendors (discussed in later chapters), and potential investors and customers alike. The result of such overenthusiasm is major loss of credibility.

Like everyone else, founders run to people they know. Given a creative idea, they go to known entities to test the concept and for validation and advice—and funding. But founders tend to go to these resources way too soon and without proper planning.

That may seem paradoxical, but one of business development's quirks is that careful planning must be invoked to decide whom to use and when. For one, the founder must carefully pick certain people, typically people not involved in the current project or a future one, to use as sounding boards. That rule is not hard and fast, but the next one is. Very early on, you must identify known people who are, or at least may be, crucial to future project development, associated either directly or indirectly with funding or other project needs. Carefully guard against involving those people in any details or requests until you are fully ready to spring your deal on them. Otherwise, unless they are close relatives or dear friends, you probably will lose them. The danger is in approaching them too soon, before you are fully prepared relative to content and professional presentation. Timing is everything; you get only one chance to make a first impression.

It may seem obvious that funding is critical, but people frequently act otherwise. Remember, many people feel that once the product is in hand, it and the project will sell themselves. Few really understand the full power and benefit of funding as we use the term. Funding is the proper acquisition of enough money at each stage of development to proceed, without any gaps in momentum, through the present activities and on to the next project phase.

Judging by real-world experience, true funding is hard to achieve. Nevertheless, most founders are markedly unprepared. While it is unfair to say that funding is easy, even when you know the facts we will convey, there is really a science to successfully securing privately funded project investment. One of our highest priorities is to equip you with every possible tool for successful and timely fulfillment of this monumental element of business development.

2.2.5 Products

Many beginning entrepreneurs think the product drives the show. While certainly important, the product typically ranks well down the list of success factors. For one, a potent management team that knows how to use professional planning to acquire needed funding can carry forth with just about any product that addresses a reasonable market.

Most projects broadly depend on the basic product idea that led to initial project formation. As with all success factors, the product usually is significant in many ways. One of the most critical reasons typically involves commercial access to one or more markets, which, in turn, implies potential sales. Potential sales drive the business plan (the written embodiment of planning).

But the product is also essential to other project areas. Product nature puts boundary conditions on reasonable forms the project can take. For example, a project that involves large products with expensive parts might better use contract manufacturing rather than in-house production, at least during early assembly. That greatly lowers setup costs and physical plant requirements and so serves to reduce needed funding substantially. Product details and the type of marketing program to be used often dictates most required funding.

Developing a new product has many facets of both a business and a technical nature. The concern here centers on how the product development process meshes with overall business evolution. (Technical details of product development are examined in Chapters 8 and 9.)

Product development usually depends heavily on another success factor realm, that of marketing. As an illustration, a certain product idea may mean entering a fiercely competitive market with resultant high producer turnover. Many people ignore reality, presuming their product is so good that it will easily displace existing players and quickly command a sizable, enduring market share. However, many competitors probably claim product superiority, too, and perhaps they are in a better position to know. Also, competitors are not oblivious to your market entry and likely will fight back hard. As we will see, it is not good for a startup company to enter an existing market with entrenched players unless your product is truly unique and you address at least one market segment that needs and appreciates your distinctive features. It is important that the various success factors interact. It is the role of planning to sort out the interplay and to develop action plans that best integrate individual success factor nuances relative to your particular product and project.

A project's product-related activities can be substantial. A firm that does all research, development, engineering, manufacturing, quality, and allied functions in-house typically has well beyond half the operation tied up in product-related endeavors. Alternatively, some firms undertake only research and development, contracting out all other product-related activities. Every conceivable mix between those two extremes can be found as well. Having your product well defined can still leave the project far from a firm hold on how best to structure a company to carry your idea forth. Our goal is to provide knowledge and insight for effective choices. Sometimes multiple choices may seem viable. However, considering all project aspects in proper perspective, there usually is one best approach. Realize that your personal as well as professional goals are in the mix. We want to transcend a technical exposé to help you make choices that balance all your dimensions, personal, professional, and social.

2.2.6 Marketing

Many people confuse marketing with sales. The difference is critical, so both terms are clearly defined here, with examples for clarification.

Marketing is the inexact science of detailed customer analysis, program design, and planned product imaging, positioning, promotion, pricing, and delivery. When you conduct market research to evaluate the need for your product idea and to identify user preferences, you are doing marketing. Likewise, when you develop plans to image the product in a certain way in each of several markets, you are doing marketing. The design and printing of promotional and advertising material is also a function of marketing, as is the act of creating pricing structures (discount schedules). You are even doing marketing when you develop transport methods and specific shipping arrangements.

Sales is a developed art form, in essence a craft, involving those activities that directly effect client purchases and service. Sales is usually considered a subset of marketing, which is how it is treated in this book. Sales is the subset of marketing that carries out only those specific activities that directly relate to actual sales events. Often, the sales function includes the customer service subgroup of activities.

Correlating previous mention of how marketing and product activities often go hand in hand, let's examine how those ideas affect marketing. What markets you wish to pursue, how you want to image and position the product in said markets, and how you price the product strongly affect the product itself. An inexpensive generic perfume in an economy pack is not likely to sell very well in high-end department stores. In contrast, a superbly crafted, precious metal key chain would be a poor fit for low-end retail racks. Usually, a high degree of coordination is required between the specific nuances of product design and the detailed marketing plan intended to move the product.

Like product-related activities, marketing endeavors can be contracted in whole or in part. Thus, one project might contract out marketing research to counteract perceived weakness while conducting all other marketing activities in-house. We will show later how many functions marked for contracting can be conducted in-house instead. Having said that, though, we remind you that the contracting out of many functions can be worthwhile, particularly when the project is relatively young. Management, working closely with the planning function, must decipher the best approach.

Attention to marketing should begin very early in a typical project's tenure. It is easy to focus excessively on developing a product idea, the result being that marketing is overlooked. However, the penalty can be severe, exemplified by projects that lock in a design approach too early, resulting in a design inappropriate for the intended market. For instance, it may be discovered later that product manufacturing costs are far too high, relative to the realistic selling price in the intended market. Alternatively, product cost may be in line, but features may be deficient relative to the desired market. The necessary alliance between marketing and product is just one illustration. All seven success factors must be integrated into one coherent plan, or the project is in serious jeopardy.

2.2.7 Location

Location is the last of the six specialized success factors, those focused on specific project attributes as opposed to global issues. The effect of location is more variable than that of other factors, and it can arise in multiple ways.

Location issues can exert significant project impact. (A detailed discussion is postponed until Chapter 14, because much relevant background and project detail have not yet been introduced.) The location issue has many facets. A list of prime elements certainly would include technological variables such as a product's development, design, prototyping, testing, production, and quality; effects of intended markets such as proximity to customers or transport considerations; specific investor demands; and risk/benefit profiles developed by looking at alternative game plans. All such considerations must be decided objectively and unselfishly by management. Additionally, there is consideration of location in its own right, that is, based on desires and constraints of the players. That should not be allowed to overwhelm other important requirements, but the desires and constraints of major team members must be given their due. It is apparent that the issue of location can become a major point of attention and even conflict.

2.3 Pulling It All Together

The goal here is to equip you with the knowledge, tools, and insight to create and progressively build a successful entrepreneurial project. Regularities exist in the process of bringing new products to market common to virtually all successful endeavors. In turn, regularities can be identified for failed projects as well, and broken projects far outnumber the successful ones. That fact should not be intimidating, because project failure follows from poor preparation by the people who run the project. Those people have not availed themselves of the broad and deep understanding of the business development process that has been amassed through regularity recognition and other extensive business development knowledge, the sum and substance of this book.

This chapter is oriented toward success. The seven identified success factors together form an interleaved set of checks and balances designed to reveal problems, usually when they are still potential rather than actual difficulties. The success factors commonly cover all three functional divisions of typical product-oriented growth companies, as follows:

- *Administrative division.* Critical functions heavily apply perspective, planning, management, funding, product, marketing, and location success factors.

- *Technology division.* Relevant functions deeply depend on perspective, planning, management, funding, and product success factors.

- *Marketing division.* Crucial functions strongly rely on perspective, planning, management, funding, and marketing success factors.

We define the boundaries of each success factor broadly, so in the limit, these monumental analysis tools cover virtually all meaningful business development issues.

Success cannot be guaranteed, because many things can go wrong. Even so, by being fully prepared for all major occurrences at every stage of business development, you place yourself in utmost control. You will also be postured to face and overcome, in an organized and professional manner, surprise obstacles as they arise. The alternative project approach is to forge ahead with a vague plan that must be constantly changed to deal with unforeseen events. The latter philosophy and method are totally unacceptable, being the basic root of failure. If there is anything approximating a guarantee in business development, it is that a poorly organized, seat-of-the-pants attitude and functional approach are highly correlated with failure.

The focus to this point and throughout the book is on the privately funded level of project, larger than the normal proprietary shop effort but smaller than the typical venture capital endeavor. Having introduced some central ideas in Chapters 1 and 2, we now attend to formally framing future discussion in Chapter 3, a concerted examination of the overall business development process. That chapter breaks down the process into unit elements, dissects each one, explores how elements interrelate, and develops the underlying rationale for the described structure. Rationale paves the way for deep and broad understanding, that perspective element we find so meritable, and embellishing your business development awareness with perspective is the highest goal we could imagine.

3

The Stages of Business Development

3.1 Overview

For discussion, the business development process is usually divided into several stages. This division artificially breaks up a process that is really continuous, but, conducted with care, it greatly aids discourse.

We are about to list various business development stages, but the labels need explanation. The stages are assigned terms that resemble phases of the product development process, which are familiar to the technical person. Still, we must stress that the stages defined here and employed throughout the book cover whole-company development, not merely its products. This book is about business development, product development being a minor subset thereof; thus, the following treatment references all administrative, technical, and marketing functions of the entire business.

Figure 3.1 and the following list break the business development process into functional stages and identify three paramount functions (planning, the business plan, and funding) transcending all business development. Each heading also reflects the primary topic for Chapters 4 through 12.

1. *Concept.* The original idea (the primordial thought) is first developed into a product concept and then into conceptions for a fully functional business.

2. *Planning.* The planning stage comprises all activities used to define, forecast, and anticipate project detail and flow in terms of time and resource requirements.

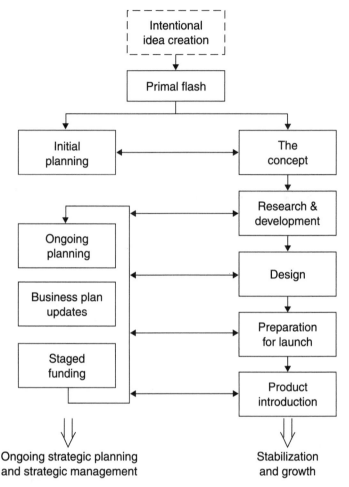

Figure 3.1 The stages of the business development process recommended in this book.

3. *The business plan.* The written business plan is a detailed, regularly updated, documented abstract of the state of companywide planning at a defined locus in time.

4. *Funding.* Funding is the ongoing process of finding, securing, and servicing needed company capital and other resources.

5. *Research and development (R&D).* This stage develops key concepts into a well-described business with proven administrative, technical, and marketing feasibility.

6. *Design stage.* In this stage, the central focus is designing and testing the product and structuring the administrative, technical, and marketing programs to build and coordinate the business.

7. *Preparation for launch.* This stage features all actions to get the product, marketing program, and business ready for initial product release.

8. *Product introduction.* This stage includes carrying all project functions from launch through the major dynamics of early production and sales.

9. *Stabilization and growth.* This stage moves beyond introduction by evolving a rational balance in companywide functions versus real-world demands.

The defined set of stages and functions is the frame within which this book develops many of its messages.

The process begins with either the intentional creation of an idea or a primal thought. The primal thought turns on initial planning activities to progressively build the concept. The concept, in turn, is the first of six sequential stages that end in stabilization and growth, which has a pronounced but indeterminate duration. (Successful companies exist beyond the stabilization and growth stage, but that discussion is outside the realm of business development.) Parallel with the core business development stages are the three critical, ongoing, recursive activities of planning, business plan creation and maintenance, and staged funding programs. Those three functions typically exhibit major interactions with most or all business development stages.

Of course, each stage will be addressed in full view of the seven success factors discussed in Chapter 2, which form a grand standard useful in assessing diverse business development considerations.

Because the central subject of this book is business development, we detail all project elements, including every aspect of the administrative, technical, and marketing functions. It is assumed that the reader has some product development experience, given the audience of engineers and allied personnel that this book targets. For completeness and to accommodate background diversity, though, we avoid assumptions of prior knowledge where feasible. The intent is to provide a complete general baseline reference for all product-oriented, high-growth business development at the privately funded tier of complexity.

3.2 The Concept

Any project is based on a concept. Academically, the concept could be defined in terms of a primal flash of insight or as the more sedate image that unfolds after studied consideration. The latter view makes more sense for the discussion here, so future references to "the concept" will mean the mature version of evolved thought that forms basal project definition.

The original idea often evolves beyond recognition as a project advances. The concept may even continue to evolve after a true project is well defined. A project's beginning is not usually finely delimited anyway. More likely, an early series of product concepts slowly grows into a broader project (business) concept, including evolved product notions but extending far beyond. The exact temporal origin of a project, therefore, is not a useful mark.

The product concept and the project (business) concept can be clearly distinguished but not easily separated. The very notion of a business (an imagined, product-oriented commercial company focused on product conception, design, production, and marketing) is heavily interrelated to the product concept (a vague mental view of a functional device or software package). From now on, therefore, an unqualified reference to the concept may involve the product or project, depending on context. The difference should not be taken seriously anyway, because the product and the project become entwined as thought evolves. Besides, our goal is to lay out a project plan that covers all bases and therefore both vantage points.

To be useful, the concept must be developed to some degree, a topic that is elaborated in Chapter 4. Of current interest is that ultimate concept results are product definition and, later, project definition. A primal flash of insight rarely reveals even an outline of a full-blown project. Such an end requires successive detailed analysis and interpretation.

In reality, concept transformation into the seeds of a project comes about through planning. However, we reserve planning to mean something more specific, as outlined in the next section.

3.3 Planning

The need for planning greatly affects every business development stage in all projects. We define planning as all activities used to specify, forecast, and anticipate project detail and flow. Planning is conducted in terms of past and present performance and future time and resource requirements.

Although *planning* is a widely used term, in this book it is used to imply the formal activity of project planning. That is meant to define all elements of the project's past, present, and future important to infer the project's dynamic behavior as a whole and the behavior of its individual functions. The process involves selecting a uniform frame of reference, almost always money, where the known past and present can be directly compared to the uncertain future. Distinct advantage accrues from using money as the frame of reference for planning. To realistically forecast cash flow (the income and outgo of money), one must seriously estimate virtually every activity of all company functions far into the future. Relative to our focus on product-oriented business development within the privately funded sphere, planning typically extends five years and therefore usually would cover all the project development stages defined in Section 3.1.

The goal of planning is to estimate and control the future. The approach is to plot all main features of company operation, in parallel, as the project is seen to advance through the chosen forecasting interval. That forces the planning process to project each company function and how the individual functions interrelate in as great a detail as possible. Each function is quantified by reducing relevant activities to a time plot of associated costs. In that way, the final summary of detailed planning is a financial spreadsheet. Inherent in the spreadsheet are a wealth of details concerning exactly how, at a given point in time, management sees the project unfolding. In addition, the spreadsheet summary may be supported by many other forms of information documenting detail and rationale.

An ultimate tool of the planner is financial modeling. It is the function of planning to pull together all defined, forecast, and anticipated ingredients of the business to build a financial model of the future. (Chapter 6 and Appendix A are comprehensive examinations of the financial modeling process.)

One key skill of sound managers is the ability to meaningfully combine and interpret mixed type of data (relative to multiple project functions). That requires substantial insight acquired from past management experience, a solid base of common sense, and thorough knowledge of specific products, markets, and businesses at least analogous if not coextensive with the immediate project. Because the project team can draw only on their direct collective experience, knowledge, and instincts, the criticality of project team makeup is underscored.

One must conclude that crucial early planning functions include defining the project and beginning to build a project (management) team. At first, perhaps only the founder is aboard, a common occurrence. The founder has a vision, refines the idea into an early take on a full project, and may even carry the work to a more advanced development stage. But at some point, almost without exception, the requisite work and skills dictate the need for help.

That juncture is critical. Experience shows substantial motivation to grab any and all willing and available assistance, supposedly solving the help problem and freeing the founder to address new tasks. That tidy picture, though, can mask impending disaster. Recruiting people based on proximity rather than on credentials and abilities is almost always bad and can be a business killer. If super talent is readily at hand, snap it up. Do not ever rush to fill slots to quickly relieve short-term obstacles. Wait for the right people but find them as fast as possible, a process developed in Chapter 5.

Planning predicts the future in part on the basis of a set of assumptions. Planning formulates premises and then uses educated guesses, published forecasts by others, and available data to model the future. A model is a mockup or mold, a simulated rendition, of expected happenings if all assumptions are true. Many view planning as a wild guess, replete with so many risks and unknowns that it is essentially useless. We will show that planning is essential and a reasonable and powerful process.

One plan may not be enough. Many times, multiple models must be run to permit comparisons. The best approach involves systematically varying the assumptions and developing a unique model for each premise set. That is one way to help gain perspective, a powerful success factor.

Planning is an iterative process. Some initial assumptions are chosen on the basis of the best available data and insight. A model of itemized company costs versus time is then built and its predictions analyzed for realism by looking, for example, at relative costs versus sales revenue for every period within the model's time span (usually several years). You subsequently note ways that the model is unrealistic, logically adjust the assumption sets in an attempt to overcome perceived problems, and run the model again. With progressive iterations, the model should approach states where the various cash flows (costs and money inflows) are balanced enough to make reasonable sense.

Once cash flows begin to balance, a model has much to say. Early in a project, all cash inflow probably derives from investment. Later, the goal is for all cash flow to come from sales revenue and for revenue to nicely exceed costs. The art of model building and analysis helps management see what costs and cash inflows might look like over time, given various combinations of administrative, technical, and marketing approaches.

3.4 The Business Plan

Planning is essential in many ways and at all stages of project tenure, early, mid-course, and later. Early planning is essential to acquire first funding and to launch efforts along an optimally productive project track. The potent influence of planning continues mid-course, due to sustained high funding needs and for dealing with critical existing and future product, marketing, and company decisions. Finally, planning retains an indispensable role in a firm's later tenure, because issues bearing major import or that affect company survival arise all the time and in many guises.

As time advances, the project needs a means to track and communicate planning. One way is through broad and frequent interaction of each management member with all or most remaining members. That usually is not enough for two reasons. First, people tend not to communicate effectively when they are very busy, and startups epitomize the term busy. Second, people tend to talk to those like themselves, so meaningful interactions between, say, accounting and engineering people or marketing and QC people rarely happen spontaneously. For those reasons and many more, the central results of planning need to be documented in written form.

A second and very effective way to track planning is the business plan. This is a major written compendium of planning that serves to collect and compile selected abstracted results of planning into one coordinated description.

Many, if not most, people new to the business development process do not see any need for the business plan, reasoning that it consumes valuable time and merely documents what is already known. While seductive, that notion is wrong, because without proper, appropriately documented planning, the project is probably all but doomed.

Although crucial, internal dissemination of information to management is only one of numerous business plan purposes. The project must exchange information with many other groups, including employees, vendors, business and legal professionals, and advertisers. For such interactions to bring about desired results, project team members must present a united front in every respect. That is certainly true relative to company employees, since mixed signals from management can precipitate insecurity, discontent, and resulting productivity problems. It is also true in interaction with suppliers of goods or services, because inconsistent interactions can hamper trust and performance. Finally, management consistency is critical for dealing with the world outside the project, such as bankers, legal people, and government agencies. Company management must be harmonious, cordial, and completely consistent over the long term. The slightest dissent that comes to the attention of people or groups outside the management team can disable a project.

The written business plan creates management consistency by providing a practical summary and interpretation of ongoing planning. The business plan is the frame of reference against which all decisions and activity must be based, with all players constantly checking their activities, thoughts, and decisions against it. The business plan serves as a reference for action and to coordinate separate functions so they act in concert for the common good of each player and the company. The plan also serves as a forecasting tool to aid funding, resource management, and company development.

Business planning must be continuous, but it is not practical to update the written plan every time there is the slightest change in some project aspect. Still, the plan cannot possibly serve its many purposes if not kept reasonably current, so the plan must be redone many times over the months and years of project startup and progressive fulfillment. The required frequency will vary with circumstance and will have to be defined and repeatedly reevaluated by management as the project advances. A rigid plan-rewriting schedule will not work either, because business needs in general and young company internals in particular are dynamic. The management team must truly understand the need for keeping the plan up to date, and must ensure that plan upgrades are in fact undertaken as changing situations dictate.

Chapter 6 and Appendix A are devoted to a comprehensive business plan treatment, eclipsing this brief introduction. Still, it is critical to know general plan content as we introduce and analyze other business development process elements. The business plan as depicted in this book has three major divisions:

- *The narrative*, the main text of the plan, often initiated with some form of summary or abstract of the global plan proper;
- *The financial model*, typically a multiyear financial spreadsheet relating estimated costs and revenue sources for each of many time periods;
- *Appendixes*, important supporting data extending the concise descriptions in the narrative or the financial model.

We will describe a special plan structure that overcomes some nearly universal problems that afflict most business plans, even those written by long-time professionals. It should be mentioned that departments can have plans too. For instance, a marketing plan, a sales plan, and a manufacturing plan are probably the most common examples of plans for individual departments or other functional company subdivisions. A departmental plan treats the subject matter in much greater depth than the business plan, that being the deciding factor of whether a separate plan is needed. The key is the creation of descriptive plans on as many organizational tiers as necessary to effect requisite operational control. Creation of specialized plans employs the same planning processes as that used for overall project planning; any differences relate merely to treatment depth.

3.5 Funding

Funding needs can affect all business development stages in some projects, and it is essential for most early stages in virtually all projects. This specialized function is typically imperative for all activities at every stage where significant resource dependence exists.

One function of planning and the business plan is to obtain funding, but that never should be viewed as the primary purpose of the plan. Therein lies a striking example of how common sense can lead one astray. The average person launching a project feels that the only real plan value is to find money. Many people write their business plans and then ignore the finished plan relative to actions and decisions, even while their projects are still searching for money. It is readily apparent to the average investor that such plans have little or no connection with project reality. Projects like those hardly ever receive much money, and none remotely approaches the funding levels required for normal project development.

Deficient appreciation of the business plan imperative is a deep pit into which even many of the wary fall. Not properly creating or using the plan is catastrophic in three glaring ways. First, ignore the plan and you might as well forget about meaningful investment money. Second, failing to use the plan relinquishes a myriad of nonfunding benefits that the plan could bring to your operation. Finally, sidestep the plan and you and your team will look foolish.

Multiple funding stages usually are necessary to carry a project through the negative cash flow era to achieve profitability. At the venture capital tier of funding, funding stages have fancy names because the process is much more formal. However, the privately funded project typically is smaller and less formal. It is instructive for us to reference short-term (less than a year) and long-term funding as two distinct project investment classes, but both types may have multiple components, depending on project nature and development plan. For illustration, a project's very first funding is short term, intended to support activities expected well within a year. Every additional round of funding is short term in the same spirit, since the money is scheduled for activities near term at the time. Long-term funding refers to funding not currently requested but known or suspected to be needed beyond the one-year short-term horizon. (Those considerations and funding in general are the central topic of Chapter 7.)

Another problem area for the inexperienced is determining how much funding is needed. We have seen countless people look for money without knowing how much they need. They answer queries from potential investors by saying the amount is not known for sure but will be known better after project commencement. Realize that an interested investor can see scads of new business plans, almost at will. Once investors sense a cavalier attitude toward funding, they usually drop the project and move on to others with sound planning and clearly defined funding needs. Appropriate planning lays out a precise list of funding needs (among many other elements) and their rationales at specific times in the future, carefully matching what investors and other financial professionals have come to expect. A business plan is written in the language of finance (funding).

Predicting the funding need profile is one thing, but trying to find the funding is another. This brief introduction defers such detail to Chapter 7.

3.6 Research and Development

Most specifically, the term *R&D* refers to activities of defining product specifics and verifying product feasibility. Defining product specifics means working out the general nature of the product sufficient for formal design engineering to begin. R&D also includes essential product feasibility verification, meaning that the qualitatively described product actually can be built and can be expected to deliver desired function. R&D represents those efforts that carry product and business development to the point suitable to start the next project stage, formal engineering design.

We apply the term *R&D* broadly, to include all project functions that occur in support of or parallel with the technical R&D functions. In a mature company, such functions likely would be termed *departments*, and R&D would be either a standalone department or a prominent subdivision of engineering. However, we are dealing with startups rather than mature businesses, and stages will be labeled according to the dominant activity typical at the point of project advancement. The discussion at hand is about the business development stage where technical R&D activities are in the forefront.

R&D activities may or may not require prior funding. If feasible, it is good to delay funding, so the project can have more for a "show and tell" for potential investors when finally necessary. In other cases, it is best to secure funding early, so R&D can advance deliberately. A crucial planning task is to estimate the timing and the amount of required funding.

R&D comprises two closely related functions, research and development. Research is a recognized function that involves combined literary and experimental methods to determine if the *product idea* is feasible and how a workable system might best be approached. Development takes the next step, to decide if the *product* is feasible. Once a product is in development, it is assumed that the product idea exhibits significant merit. That does not automatically mean that a product with proper features can be suitably made at reasonable cost for a particular market under expected marketing cost/benefit assumptions. Development's job is determining the latter collective of complex issues.

Actually, both research and development involve, in one way or another, all project functions. We have depicted R&D's center stage as occupied by the technology realm, but administration and marketing also are critical. Administration provides the environment for project existence, being responsible for original project formation, directing all planning and the business plan, and providing productive space (physical or virtual) where project activities occur. In turn, marketing is deeply involved in both research and development, because activities conducted by administration and technology depend in part on marketing-centered planning, knowledge, data, and wisdom.

The same joint effort between administration, technology, and marketing is involved in the concept stage, planning, the business plan, and funding. To some degree, administration, technology, and marketing become involved almost immediately after the initial flash of an idea, but the nontechnical functions can get slighted because of early product overemphasis.

The purpose of the R&D stage is both narrow and broad. In the narrow view, R&D verifies feasibility of both the product idea and the product. More broadly, R&D further builds and validates the entire business in preparation for future development stages, most immediately the stage highlighting engineering design. A detailed discussion of the entire R&D stage is in Chapter 8.

3.7 Design Stage

The most prominent project function is product design (detailed engineering design). Detailed design usually requires some funding to be in place. However, given the book's engineering-oriented audience, it is possible that a project might advance into or beyond detailed design without requiring outside capital.

Detailed design involves sequential steps that progressively define product detail. The process begins with what we term a *marketing specification* (marketing spec), a detailed product description cast in the language of the ultimate user. As an illustration, some marketing spec items for a pencil might be a small, lightweight, very low cost, hand-held device, easy to produce, with access to a renewable writing function on one convenient aspect and an eraser at another.

The marketing spec can be viewed as either the final R&D act or the first step of detailed design. The latter view is preferred here to better correlate with other specs introduced in the material to follow. Regardless of academic placement, creating the marketing spec is a monumental beachhead led by the marketing function. Technology also must be involved to keep technical issues honest, and even administration must participate to ensure proper planning and forecasting interplay.

The marketing spec imposes constraints on detailed design, but it is important that it be specified in the language of product users, not designers. Engineering should be allowed significant latitude to create myriad designs, all completely consistent with the marketing spec.

Marketing spec translation into design language results in what is termed the *design spec*. Technology leads the translation but must draw on strong marketing input. An engineer might ask, "By *lightweight*, do you marketers mean less than a pound, less than an ounce, or what? Way less than an ounce? OK, can we agree on less than or equal to half an ounce? Done." The same process continues for all other items in the marketing spec. Marketers are considered the experts on user desires and needs, whereas the engineers represent technical considerations. Together they develop an initial design spec that fully represents in design language the marketing spec framed in user language.

The design team then develops early design concepts, based on the initial design spec. Design and marketing should interact again to review and refine design concepts, resulting in an updated design spec. The process of give and take between design and marketing may occur several times, resulting in added design spec updates. Once the iterative process settles on a unique design concept, the design team defines complete product detail documented as the detailed engineering design. Then joint meetings with marketing should alternate with further detailed design until both groups agree that the design is complete. Certain stages in the iterative process may require physical models or prototypes to answer questions of feasibility, materials, cost, weight, user preferences, and many other issues. The entire detailed design process is the subject of Chapter 9.

For our purposes, another function of the design stage technology division functions is solving initial production and quality issues. As the detailed design unfolds, it is critical to begin addressing exactly how and where the product will be produced. The issues of parts and materials purchasing, product packaging, shipping, transport, returned goods, and other production-related issues are important as well, and all results should be documented in a *manufacturing spec*. Finally, the entire design process must provide significant attention to quality considerations. Set forth in a *quality spec* are all the quality requirements to be imposed on design and production.

Marketing has much more to do than the design contributions mentioned previously, including initial planning for product introduction and assuring marketing needs are properly recognized in administrative plans. Administration is quite active too, primarily concerned with funding, planning, updating the business plan, and managing all resource needs for current and planned activities. The spotlight is on technology, but strong contributions by administration and marketing are paramount.

3.8 Preparation for Launch Stage

The design stage must fully prepare the company for the subsequent stage, preparation for launch (or, simply, preparation). Preparation requires extra management team effort to pull all company aspects together, an ongoing charge that is especially intense due to impending product release for sale.

Notice that the preceding paragraph referred to the *project* as the *company*, the approach to be taken from now on, to emphasize that the project at some point will take on some type of formal structure. In fact, one would expect a formal structure to be in place by the preparation stage (most likely before), to properly recruit and assume funding. Parenthetically, a company is not the only formal structure the project might adopt, just by far the most useful and most likely. All types of potential project structure and the rationale for each are discussed in Chapter 5.

The preparation stage challenges the three functional company divisions with a unique constellation of issues. The main thrust of administration during the preparation stage is, minimally, to ensure appropriate resource availability. First, it is important to evaluate and adjust management team capabilities to contend with impending company moves from a sheltered environment to one exposed to many new outside forces (e.g., competition, legal exposure, insurance needs, customer and public scrutiny, increased personnel issues). Second, ongoing planning must be expanded to account for the previously cited broader, real-world interaction. Third, the business plan must be updated whenever necessary to reflect changes that dynamic events and altered planning probably will compel. Fourth, proper funding must be secured to support the various activities of the preparation stage.

Turning to technology, the preparation function must complete the core product design and clean up various design-related tasks such as product or test documentation, completion of ongoing lifetime tests, creation of manuals, development of returned-goods procedures, and so forth. In addition, the whole issue of manufacturing must be addressed, only a small part of which likely was carried out as part of engineering design. Finally, the preparation stage interval can be important to refine quality system philosophy and to develop and implement complete quality programs and procedures.

Preparation is an especially dynamic time for the marketers. With actual product launch impending, this stage is when marketing must finalize all program design, customer analyses, and plans for product imaging, packaging, promotion, distribution, and pricing. The sales program also must be finalized, as must the customer service function, and early actions under both functional categories may begin during preparation. For instance, some early marketing may be planned prior to product launch, to build name recognition and to begin creating perceived need on the part of potential customers. The activities and positions adopted by the marketing division during preparation can be crucial, because first impressions become expended, and it may prove difficult to change course later.

The three functions of administration, technology, and marketing were treated singly, but comprehensive coordination among these essential functional divisions is paramount. Coordination is a major goal of planning and the ultimate responsibility of administration at the company level, but it is critical for all functions at the divisional level, too.

3.9 The Product Introduction Stage

In part, a given stage of business development must properly position the company for the next development stage. Above all else, preparation must fully ready the company for the subsequent product introduction (or, simply, introduction) stage. Introduction is the big event for most team members, often viewed as what all the effort has been about. Still, it is just another stage of the overall business development process. Following is a brief look at major concerns of the primary company functional divisions during introduction.

Administration must be at the height of readiness. Early detection and rapid response may be necessary to contend with all sorts of unexpected dynamics, such as shifts in actual inflows and outlays versus budget or surprise moves by competitors. Administration also must, as always, create an optimal environment within which technology and marketing can flourish. Planning is the watchword, and the written business plan must be kept absolutely current, because, with little notice, the company may have to seek funding, negotiate new or revised arrangements with vendors, complete new agreements with major customers, or carry out a host of other short-term tasks that demand a current written business plan. The company must posture to avail itself of every possible opportunity and effectively respond to the unexpected.

Introduction challenges technology in part through initial, real-world product exposure and myriad potential problems not anticipated during product design. It is critical to mount a sound but rapid design fix to any and all requisite problems. Many types of manufacturing problems may arise as well, often as critical as design imperfections and demanding immediate attention to minimize product flow interruptions. All problems of a design or production nature trigger quality responses, too, and all technology functions must be carefully coordinated.

Last, but certainly not least, is the marketing function. The introduction stage for marketing is its coming-out party, the moment of truth. All past marketing research, program design, and product management activities may seem irrelevant (even though they are not), because suddenly assumptions are upstaged by real-world scrutiny on a grand scale. It takes deep experience, strength, and discipline to forge ahead with prearranged plans, in spite of the inevitable critical responses. Adopt the posture that only compelling reasons will be allowed to induce changes in carefully considered plans that, according to our business development recipe, will be in place.

One momentous, companywide issue at the introduction stage is initial dealings with the public. Positive reactions certainly are welcome and even used in promotion, while negative reactions must be addressed quickly and effectively to minimize their impact. Managing outside distractions competes with in-house governance but is critical for company control and advancement. Finally, the biggest hurdle of all is keeping abreast of resource requirement changes associated with high company dynamics, and managing such needs to mitigate short-term problems without disrupting longer range plans.

As before, tight coordination among the three company functional divisions is paramount. Probably nothing like the dynamic world of introduction has been encountered before. The situation is manageable, though, given proper preparation.

The introduction stage of business development lasts much longer than the benchmark day of product release for sale. Introduction spans an interval beginning just prior to product release and ends, by our account, when the initial promotional activities that both label and exploit the product as newly released have run their course.

Usually an introduction stage also will include the peak or passing of several other key events or processes. By way of illustration, many early product or manufacturing problems, packaging and shipping issues, and interfunction communications glitches will have arisen and should have been reasonably resolved. Introduction is a challenging test of the planning and other general management plans and methods previously installed. Finally, somewhat akin to the preparation stage, the time course of cash flow through introduction is a good measure of financial prediction and control methods. Once introduction has been accomplished, company dynamics usually subside somewhat.

Easing of pressure in late introduction provides an excellent opportunity to incorporate lessons of recent activities. This stage provides a tremendous learning ground because of high dynamics, and wise managers quickly work such lessons into company functions at all levels, an axiom to be honored at all other stages too.

3.10 Stabilization and Growth

Introduction typically is followed by a period in which the dynamics of change slow down a bit. The crises of introduction have come and gone, and new problems usually are not so frequent or so traumatic. For broad applicability, we limit reference and discussion to the initial product, but a given company may have more new products lining up for launch and therefore may not see much easing of dynamics.

The stage immediately following the introduction stage is labeled stabilization, to highlight the period's usefulness in cleaning up loose ends and mending wayward processes. After stabilization is growth, a time for consolidation of individual functions and the business, emphasizing market penetration and production cost reductions. A comprehensive discussion of stabilization and growth is in Chapter 12.

3.11 Summary

In summary, this chapter began by segregating the business development process, each stage being named according to the company function most prominent during the subject period. All company functions, however, are important at every evolutionary stage. At the very beginning, there may only be one person, the founder, who conducts all functions. But as the initial product idea grows into a project, and the project progressively evolves into a company, more and more key players typically are required. Should that not happen, logical professional progress toward a real company may not be project destiny. Usually well before product introduction, the project has become a company and the company has begun to assume a life of its own.

This chapter briefly reviewed each business development stage to provide orientation and familiarity before the more detailed treatment to follow. A similar technique is used throughout the book, a particular concept being revisited many times but from a different vantage point, level of detail, combination relative to other concepts, or other distinct presentation. Facts are facts, but an appreciation of how a web of facts is interrelated underlies real wisdom.

We have now introduced all business development stages, plus much background information, to build a sense for how the process works and why. Chapter 1 was a whirlwind glance at business development, product ideas and their evaluation, types of development projects and a focus on the privately funded realm, the seriousness of entrepreneurship and requisite commitment, and the type of person best suited for this type of endeavor. Chapter 2 introduced and significantly developed seven success factors, key attributes common to virtually all winning projects. Also discussed was failure and regularities appearing in projects that ultimately fail. Finally, Chapter 3 examined introductory structure and content of each business development stage treated relative to project advancement.

Part II breaks from the sequential course of this chapter to cover some key ancillary material urgent to our story. These imperatives, conceptualization, planning, the business plan, and funding, strike at the very heart of countless details and rationale that must be addressed concerning company definition, structure, and development. Once Part II has been completed, the step-by-step tour of the business development process resumes in Part III.

Part II
Professional Planning and Funding

All facets of business development are demanding. That means each of the seven success factors developed in Part I represents a realm of difficult challenges for the uninitiated. And hardly anyone would disagree that conceptualization, planning, the business plan, and funding are the most daunting of all.

Although still tough, your situation should be better than most. Like everyone else, you must pay reverence to the seven success factors, that being requisite to business development success. That means comprehensive, meticulous, and objective treatment of diverse issues, because the results will be given keen business scrutiny. Even so, you will have a supreme advantage, if you adhere to this book's prescriptions and know the means to be fully prepared.

The next four chapters comprise Part II. Chapter 4 thoroughly examines the original product concept. We show how the primordial concept grows from a narrow idea about a product, developing over time to become first a more detailed product idea and eventually expanding to include an accompanying project (business) concept.

Concept evolution is a type of planning, and the discussion in Chapter 4 introduces basic principles of this critical discipline. Chapter 5 then develops planning more formally. In one sense, planning is the key to success, because project fulfillment can come only with relentless, concerted, professional implementation of winning action plans. The most visible planning embodiment is usually a company financial spreadsheet or model, and discourse centers on how to build such a model and conduct other major facets of planning in an organized and professional manner. Primary topics include team building, research, strategic and tactical planning, and formal organization of the business.

Prior treatment of planning sets the stage for examining the written business plan in Chapter 6. The business plan is carefully dissected to cover the rationale, structure, content, and style of this imperative document. Additional issues include alternative or contingency plan development, and how the company and its intellectual property can best be protected long term. The business plan must be current for each required use, so many updates are obligatory over the extended business development process.

Finally, Part II addresses the overriding matter of funding in Chapter 7, essential to ensure proper, timely resources, and shows how this gargantuan barrier for most projects (those of the unprepared) can be tamed. Fully accommodate all seven of the success factors introduced in Part I, and the project will command powerful advantage. You will distinctly outshine competitors (other projects competing for the same money) in substance and style.

4

The Concept

4.1 Just What Is a *Concept*?

The notion of *concept*, as developed throughout Chapter 3, resembles a process more than an event. Briefly, the concept is an extended process that begins with a singular flash of insight, the "Aha!" phenomenon, or what we term the *primal flash*. The primal flash often quickly dissolves through further thought in a flurry of concept extension and extrapolation, successive transformations leading to altered and more enduring concepts predicated on but perhaps quite distinct from the primal flash.

Entrepreneurs can usually remember their primal flash weeks, months, even years later. The primal flash is a true, one-time, memorable event that is incessantly subjected to modifying scrutiny and analysis. Good ideas continue to command attention, at least in the background, for a long time. Ongoing critical thinking usually leads to progressive deviations from primal flash details over time. In the limit, comparing the primal flash to a complex concept having evolved therefrom can be like night and day.

Concept transformation away from the primal flash is not detrimental, because successive rounds of review and alteration are natural, good, even necessary. In fact, a nonevolving idea likely is not being pursued or is just not much of a breakthrough after all.

The primal flash usually represents a mind's eye view of a new product or, one step removed, seeing a need and then a product to meet it. Either way, the primal flash is product dependent. Even so, somewhere along the trajectory of thoughts, the product concept picks up an accompanying business concept, some at first vague notion for product commercialization.

The appearance of distinct product and business concepts is important, signifying ongoing active mental attention to the evolving concept. Right now, if challenged, you probably could create or recall a few quick "ideas" for potential products (e.g., a broad-beamed laser to kill insects, sort of a laser shotgun). Most such notions do not withstand casual analysis (e.g., the excessive power requirements of the laser shotgun) and are soon forgotten. An idea that endures such normal mental processes and still has meaning and significance is a candidate for further pursuit.

We have found that many budding entrepreneurs do not understand idea evolution or management. Concepts must be tested against reality and successively groomed and retested perhaps many times, to develop productive renditions practical for promotion to formal product and business development. The earliest prominent phase of a project we have labeled the *concept stage*, signifying the prominent focus on concept development in that interval. This chapter involves activities that occur primarily during the concept stage. Even so, the discussion basically relates to the concept proper and its development, that being in essence the only activity of the concept stage proper.

The concept functions as the seed from which all else grows, the spring board to product and business. An authentic, developed concept, rather than the primal flash, drives project origin. Beyond that, concept nature can place substantial constraints on practical project development. For instance, a concept that depends on the product floating in water restricts average product density, and invasive medical products must be sterile. Remaining chapter sections carefully build a concept development approach, said to occupy the concept stage of business development, featuring repeated tests of concept feasibility designed to reveal problems early and provide opportunity for corrective action. Ideas should be verified, then fixed or forgotten early, to avoid waste and disappointment.

4.2 Creating a Concept From Scratch

Say you want to start and build a business, but a suitable idea has not yet appeared. That is not a major obstacle. Concepts can be created from scratch. In fact, that can be a nice way to proceed. By controlling the idea creation process, you can significantly influence the result. In other words, you can often work backward, selecting a general industry or niche and actively create an idea targeted to that realm.

This section discusses two ways of creating "designer ideas": (1) a takeoff on what is called market segmentation and (2) brainstorming.

4.2.1 Market Segmentation

Market segmentation means analyzing an entire market to identify its component segments. The process usually is used in conjunction with what is called product positioning, in designing detailed marketing strategy (see R. F. Smith's *Entrepreneur's Marketing Guide*). We adapt the technique for use in a novel way, finding gaps in either an existing product spectrum or in meeting user needs that suggest a product opportunity.

An adapted segmentation analysis of a given market based on existing products begins by laying out those products in a list or diagram. List all significant device attributes (e.g., price, feature complement, uses, physical parameters, appearance, quality and reliability, repair and maintenance, manuals, return policy, warranty) in a tabular comparison of all models from each manufacturer. Finally, assess the degree to which the entire retinue of product attributes covers the spectrum. Weak or missing areas are gaps to be evaluated for new product opportunities, which are treated thereafter just like a primal flash.

In turn, one step removed from a product approach, a segmentation analysis also can begin by a listing of all recognized needs served by an industry. One way is to use promotional literature from all known products within the relevant market. Even discontinued or historical pieces can prove to be valuable. Written resources about industry user needs, found in scientific and technical journals, reference materials like the Thomas Register®, or back issues of trade magazines, also can be helpful. (Keep in mind that many such sources talk about user needs in the act of trying to create demand, that is, they are promotional pitches.) Analyze the resulting real-world user need list, searching between the items for modern needs that are not being properly serviced. Finally, think about unique potential products that could fulfill the identified unaddressed needs and treat those product ideas just like any original idea or primal flash.

4.2.2 Brainstorming

An alternative to market segmentation, brainstorming is a simple, effective approach that is applicable far beyond the current discussion of controlled creation of product concepts. Structure and leadership are essential for effective results properly coordinated with company mission and strategies.

Like segmentation analysis, brainstorming can be utilized with a plan of attack that is based on either an existing product or an existing user need. First, define the general product or user need market of interest, carefully establishing the "width" of the analysis window. Window widths that are too wide or too narrow create excessive or insufficient data, respectively, and some experimentation may be in order to optimize resolution. Next, build a list of every possible idea or comment of slightest relevance to products or user needs within the study window; historical, existing, hypothetical, and even seemingly nonsensical notions are equally useful. Try combining or splitting items as well, even if relevance is not obvious. Nonsensical or otherwise seemingly irrelevant items can suddenly become relevant upon further analysis or trigger ideas that are.

Brainstorming can be done alone, but using multiple participants is better. Display the list on a blackboard, flip chart, overhead projector, or other medium where everyone can see the full record. Schedule three to four sessions of several hours each spanning not more than several days. The breaks aid fresh thought and the limited duration enhances coordinated, collective perspective. Interplay effectiveness falls off rapidly when the intersession interval extends beyond a day or two.

The list should be allowed to grow for as long as those involved feel useful material is forthcoming. As the sessions continue, the appearance of new items will decline, and the process must stop sometime. The biggest risk is in stopping too soon, so proceed until those involved feel that novel ideas have run their course.

The same group should hone the list until all that remains is a core group of top ideas. One would expect a substantial reduction of the list, compared to the full roster, but realistic limits depend on the niche studied. Hone the list again: Divide items, rearrange and recombine them, discuss each item and the interrelations to see if new thoughts appear—in other words, manipulate the list in every conceivable way to pull out novel elements. Then hone the resulting list one last time, creating a final roster of either product ideas or user-need ideas, whichever was the goal. Analyze every entry on the final list to pick the one or more items that make major business, technical, and marketing sense on first analysis. Take a last look at each item to see if there is any further way to mix, combine, or divide items in a productive way. Finally, select the one or more product brainstorming concepts that best meet your goals.

To this point, the focus has been on several ways to create fresh product ideas, both directly and indirectly. Once an idea exists, it is irrelevant whether the source was a primal flash or an orchestrated process, or if the product idea was approached directly or arose indirectly by assessment of a user need. Raw ideas, independent of origin, must be subjected to a feasibility analysis, the topic of Section 4.3.

Save all records of the various analysis stages, including supporting material such as brochures from existing products, research results, and crude notes. Months or years later, company thought about new products or user needs can greatly benefit from such historical material.

We caution you not to introduce historical lists too early during a future analysis. First, conduct the whole process anew, from the ground up, without consulting previous work. In fact, it can be valuable over a series of major analyses to use distinct approaches. For instance, cycle between brainstorming and segmentation analysis or between user-needs evaluation and product analysis. During any one major analysis program, after most or all new ideas supposedly are on the table, introduce lists from previous major analyses to see if any further thoughts are triggered.

4.3 Concept Feasibility

Now you have created a concept that originated either as a primal flash or from intentional creation. Most likely, you have further reviewed the idea mentally to see if it makes sense, at least superficially. Still, it is just an idea, a concept perhaps developed mildly beyond the primal flash but untested in any formal way.

Further concept development can be arduous, as discussed in Section 4.4. Prior to extensive development, it is sensible to test concept feasibility, the concept being fresh enough that a quick, deeper look may reveal impractical or impossible elements.

An analytical frame of reference is needed for the initial feasibility assessment, and we have chosen the familiar three functional divisions of the company: administration, technology, and marketing. That helps interweave the notions of company and product concept in a natural fashion.

We now examine the concept relative to each respective company division, a preliminary process that is repeated later in a more rigorous fashion. The goal is to confirm (or deny) sufficient product feasibility to justify the ensuing major step discussed in Section 4.4, that of concept development.

4.3.1 Administrative Feasibility

Early concerns are to build reasonable confidence that the product concept could support a sound business. The exact nature of the business is not known, pending later detailed planning, but the issue here is more general, whether a profitable business offering the concept product is at all plausible. We leverage the fact that the nature of competing products and their producers are known or at least can be developed.

The job, therefore, reduces to a preliminary evaluation of whether the marketplace can viably incorporate another producer, an administrative rather than a marketing analysis of whether to form and build a company at all. The intent is to identify any major financial, economic, regulatory, demographic, cultural, or social barrier to market entry and company success.

A useful yardstick is a "venture capitalist test," a mental aid to perspective that can be an invaluable qualitative planning tool helpful in resolving otherwise refractory decisions. It works like this. Let's say, qualitatively, you wonder if there is a market for a newly conceived product. You have some crude sales estimates and competing product literature acquired from industry equipment users and product content ideas based on the same discussions, but weighing various elements proves difficult. Imagine standing in front of an experienced venture capitalist, making a verbal presentation about your product concept and how it stacks up in the marketplace. The test is whether you would feel comfortable with your story, given this challenging environment. In other words, in attempting to evaluate personal thoughts, the standard should not be convincing yourself. Actually speak out loud to the imaginary venture capitalist and listen to what you have to say. Objectively done, this exercise can be quite instructive.

You may feel the venture capitalist test is not a fair one, the imaginary listener representing a real professional who knows all about business, including startups. Besides, such people usually inhabit the more rigorous realm of venture capital projects, not the middle ground of privately funded endeavor.

On the surface, that is true. The reality, however, is that if you cannot stand your ground in front of a ghostly venture capitalist, you also will fold before the investors of the privately funded world. When it comes to being able to address the tough issues (e.g., market size and turnover, applicable sales channels and buying methods, price versus production cost, labor and facilities availability, project attractiveness to investors), there is scant distinction between the venture capital tier and the privately funded project tier. The prime difference is the amount of money, although there are others. In either realm, though, money is almost impossible to find, with one exception. Those projects that have taken great care to accommodate all seven success factors (not most, but *all*) suddenly gain exposure to real money.

4.3.2 Technical Feasibility

If the concept passes early administrative muster, it then should be tested in the technology division's arena. The types of questions here do not necessarily require extensive research, although some study may be advantageous. Again, the goal is to ferret out any major hurdles that might provide substantial or insurmountable technology obstacles. The central issues are R&D, design, manufacture, and quality. First, look for reasons why the product could not be developed, designed, or produced with satisfactory quality. Check whether your current predesign notions of detailed design compel exotic or costly materials, technologies, expertise, processes, or regulatory compliance that might render the effort improbable or impossible. Additionally, examine whether the design can be properly protected, in a way that competitors cannot adopt the same or similar design (Chapter 6 has much more to say on product protection).

Again, using the fictitious venture capitalist, imagine some of the tough questions a thorough and blunt professional might ask. Describe how you envision the product design. Details are not yet clear, but you should be able to render a general description, such as the basic size, weight, materials, rough estimate of the ratio of production cost versus selling price, and perhaps a number of features. Now, defend the product attributes with rationale, explaining why the particular choices were made compared to possible alternatives. Then repeat the entire question series, successively replacing design, in turn, with different prime technology subfunctions, such as production, testing, packaging, shipping, and QC. Thoroughly test your current ideas in the concept development process against those questions. You may lack many answers, and some questions may be more appropriate later in business development. Upon finishing this book, you should have good instincts for judging when particular questions are most useful. Even so, confronting key questions early can build perspective and orient thought in preparation for later challenges.

4.3.3 Marketing Feasibility

Finally, examine the product concept relative to preliminary marketing feasibility. The most appropriate issues at the concept stage probably include how well user needs are met, pricing, the nature of the envisioned market(s) (or how buyers would be accessed), and sales channels (e.g., direct sales force, direct mail, catalog sales). For some products, many of those notions may be just glittering words at this point; detailed concepts await later discussion, especially in Chapters 10 and 11.

The imaginary venture capitalist can yet again lend a hand. Explain out loud how customer profiles will be developed, marketing programs designed, and the conceived product imaged, positioned, priced, promoted, sold, and distributed. The presentation cannot be highly specific, because detailed planning is yet to come. Still, the marketing function should begin developing renditions of such ideas early, to permit extended analysis and preparation. The goal is to examine elements of marketing with the idea of finding flaws in the marketability of the product concept. Evaluate all the subfunctions to see if any barriers to marketability seem apparent. This helps flag problem areas for meticulous study later and can reveal an unmarketable product concept early in development.

4.4   Development of the Concept

Assume your idea has survived the preliminary feasibility tests described in Section 4.3. The product concept has been deemed feasible, to this point in the concept stage of business development, when tested against broad, albeit preliminary, features of administrative, technology, and marketing realities.

Nevertheless, you still have only a concept related to a product, but to presage longer term activity and analysis a concept for a business is needed. Basically, you must develop some confidence that the envisioned product can carry the project through multiple business development stages toward a profitable, ongoing business. The product concept is not lost but becomes embedded in a much grander concept for a compatible business.

For generality, assume your product enters an existing market rather than opening a new one. Key parameters include market size for the product, the market share your product might progressively capture, and the product price required for such accomplishment. Other central parameters are anticipated costs to sell the product and the direct costs to make it. Also required are estimates for the costs of administration, technology, and marketing required to create the environment in which sales and production occur, including estimating promotion and production costs relative to the projected time course of resultant sales. The list could go on and on.

An additional consideration affecting the entire list is that all noted parameters interact with at least some others, so hardly any factor can be considered in isolation. Parameter interaction is an important topic itself and is best handled indirectly as a part of the discussion of planning in Chapter 5.

We now detail a number of key parameters, Figure 4.1 indicating some qualitative parameter interrelations.

4.4.1   Market and Market Size

A product's market depends on the relevant buyer population (market size) and the useful life of the product. The first parameter, the relevant customer base, is that group of purchasing entities that have any reasonable use for the product. An entity can be an individual person, a group of people, an organization, or any mix thereof. The second factor, the product's useful life, represents the average interval of expected product use, either before replacement or loss of useful appeal. The market is defined as being equal to the market size divided by the product's useful life.

For example, let's say that nuclear golf carts have an expected life of 10 years before replacement. Further assume that there are an estimated 10,000 U.S. entities believed to represent the ultimate potential purchasing universe

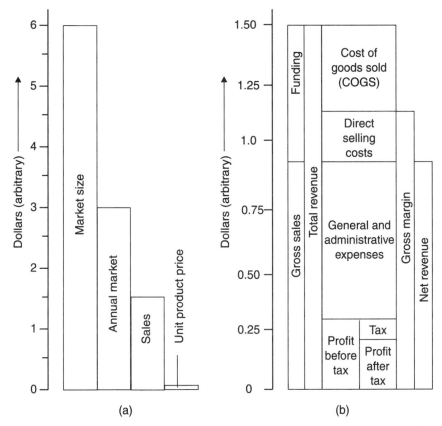

Figure 4.1 Qualitative relations between key financial terms. The two histograms depict (a) a hypothetical market with sales, annual market, and market size related 1:2:4 and (b) a higher resolution version of the same vertical axis to show relative financial term relations for the company whose annual sales are illustrated in (a). The second histogram graphically depicts that any gross sales shortfall relative to total revenue requirements must be made up with investment funding.

(those having both the capability and the propensity for purchase) for such an upscale, novelty product, that is, the market size for nuclear golf carts. It would not be useful to include people who could not afford the carts, since they do not contribute to knowledge about potential or real nuclear golf cart sales. The market for the product is the 10,000-unit market size divided by the 10-year expected product life, or 1,000 units/year.

4.4.2   Product Price and Market Share

The product will carry a price set in the planning process by company management. Product price affects sales inversely, because raising the price typically decreases sales, but the relation is not linear. It takes some thought and perhaps marketing research to develop a feel for the price-versus-share relation. Market share is simply a percentage comparison of actual sales to the total sales one would expect if every eligible entity bought the product. We have a product with a market size of 10,000 units and a market of 1,000 units/year. A hundred sales in a given year would equal a market share of 10% (100 units/year sold in a market of 1,000 units/year).

4.4.3   Gross Sales

Gross sales simply means the sum of all sales receipts in dollars, often termed gross, or total, revenue. Usually no adjustments for expenses or other factors are applied directly to revenue values before they are reported as gross revenue. The same is true for total revenue estimates created during planning.

4.4.4   Cost of Goods Sold

The *cost of goods sold* (COGS) expense group is frequently shown as a separate category on the spreadsheet, as shown in Figure 4.2, to ensure visibility, easy access, and ready comparison to revenue. COGS highlights the direct costs that arise from making the product, commonly including parts, materials, supplies, labor, and packaging. COGS may also include shipping, returned goods costs, and other ancillary functions. Each company should define an appropriate COGS makeup for financial reporting, to aid analysis and understanding.

Another term widely appearing in writings on business and business development is *gross margin*, which simply is total revenue minus COGS (see Figure 4.1).

4.4.5   Selling Costs

The term *selling costs* denotes the direct costs of selling the product. Later, you will see the idea of direct costs showing up in other contexts as well, such as design and manufacturing. The direct costs of any function are those expenses that occur in an immediate way in carrying out the subject activity. Direct selling costs reference all expenses attributable to acts of selling, such as sales commissions, travel and communications expenses, entertainment, sample product, and selling aids (e.g., computers, video display equipment, business materials and accessories).

Item	Year 1 Q1 Q2 Q3 Q4	Year 2 Q1 Q2 Q3 Q4	. . . .	Annual totals Yr1 Yr2 Yr3 Yr4 Yr5
Revenue				
Total				
Cost of goods sold				
Total				
Gross margin				
Direct selling costs				
Total				
General and administrative expenses				
Total				
Profit and loss			. . . .	
Profit before tax				
Tax at ___%				
Profit after tax				
Retained earnings				

Figure 4.2 Generic spreadsheet structure (see also Fig. 6.1(d)). Understanding of a spreadsheet is improved when related rows are collected into major groups (the five shown). The columns represent specific time intervals, quarters for the core spreadsheet and single years for the five summary columns at the far right. Four line items appear under the "Profit and loss" group. "Profit before tax" is the algebraic sum of all revenue minus all expenses. The next line item is taxes; consult an accountant for a value appropriate for you. "Profit after tax" is the amount remaining after subtracting the estimated tax amount. Finally, "Retained earnings" represents the cumulative profit after tax (retained earnings are also directly transferred to the balance sheet).

Because sales are imperative, direct selling costs often are singled out as a distinct category on the spreadsheet, being placed immediately below the COGS category (see Figure 4.2). Total revenues and selling costs are related, because, all else being equal, revenue would be expected to rise as selling costs go up. That assumes, though, that selling cost increases are due to a commensurate increase in productive sales activity, not a free-spending salesperson or increased intensity of inferior sales methods. Thus, the relation between revenue and direct sales costs is often not proportionate, and further, actual sales may lag behind associated selling costs, sometimes markedly. The relation between revenue and direct sales costs can be complicated, and a distinct spreadsheet category for direct selling costs, as depicted in Figure 4.2, can facilitate analysis and understanding.

For us, what is termed *net revenue* equals gross revenue minus COGS minus direct selling costs (see Figure 4.2).

4.4.6 General and Administrative Expenses

Gross revenue must support not only COGS and direct selling costs but all other company costs as well. These costs often are lumped as one category, general and administrative (G&A) expenses.

G&A costs include all administrative functions, all marketing activities not included in the direct sales category, and all non-COGS technology endeavors, including R&D. The aggregate of these costs can be considerable. Their collective presentation and lower spreadsheet placement do not signal insignificance—G&A activities and costs are as critical as anything else. It is convention to deal with G&A expenses this way.

4.4.7 Perspective

This chapter has introduced a number of standard terms involved in company finance. These parameters are widely employed, and good insight regarding their meaning and interactions is essential. Of course, it is important to ensure strong finance and accounting expertise on the management team, to optimize the present assessment and ensure proper company financial system design and operation as the company develops.

For now, you must assess whether the envisioned product can carry the project through multiple business development stages toward a profitable ongoing business. Administration's planning function plots the timing and the amount of funding needs, analyzing every company function at all business stages, present and future. The object is to estimate the magnitude of overall effort about to be assumed, stage by stage. Even though many details are unknown and you may sense much guesswork, the effort is of paramount importance.

Transforming the product concept into a business concept is the job of planning. Planning is the primary topic of Chapter 5, which covers the details and intricacies of how planning is done. The following material is more concerned with the results of planning.

To formally develop a business concept, you will create a detailed model (pro forma) of the business, as described in Chapter 5. Briefly, a model is a financial spreadsheet that represents not just the product but an entire business that makes the product. The model—in essence, a large table of columns and rows—shows estimated revenues and costs along the vertical dimension as a function of time depicted on the horizontal axis.

At this point, precise costs are unknown, because none of the previously introduced financial parameters have been quantitatively developed. Even so, an initial, admittedly vague model is essential at this point, estimating various revenues and costs over time. The approach is briefly as follows (it is described in detail in Section 5.3). Assume the business will be profitable in the future when sales volumes are good and costs have stabilized. Also accept that there likely will be no revenue for quite some time, because all R&D, design, production setup, marketing, some initial production, and a host of other activities must occur prior to delivering the first product. That concomitantly means expending major money and time before any sales revenue arrives. It is obvious that much investment money is obligatory to support the company until sales revenues are able to do so. That is the general reality with which almost all projects must contend.

The first step in modeling is to begin to build the spreadsheet, a matrix of intersecting columns and rows, generically depicted in Figure 4.2. Columns represent intervals of time, increasing to the right. Rows are collected into several prominent groups, as shown in Figure 4.2. The bottommost group, "Profit and loss," begins with the line item "Profit before taxes," which is the algebraic sum of all revenue (the topmost spreadsheet group) minus all costs (all cost/expense groups).

Consider the first period of time, a column of cells typically representing a specific month or quarter. Enter an estimated value for each cell in the column, beginning with the revenue element rows at the top (probably zero for early time periods) and proceeding to address each lower cell representing costs/expenses until all respective column data items have been treated. Then proceed to the next column, representing the next month or quarter, and again estimate each revenue and cost item for all relevant cells in the column. Continue building columns to the right, representing successive future time periods, to cover a time span of several years (this book assumes 5 years, the professional consensus). This brief overview is purely to support present discussion; detailed modeling is presented in Chapters 5 and 6 and in Appendix A.

The value entered in a particular cell commonly is determined in one of three ways:

- *Determination.* Some values can be accurately defined according to known agreements or rates, such as facilities and equipment leases, professional retainers, and, often, utilities.

- *Estimation.* Activities like payroll management, accounting, and tax preparation occur at fixed hourly rates but in uncertain amounts.

- *Proportion.* COGS and direct sales costs frequently are estimated better as a percentage of sales rather than as an absolute number. Actual future sales are estimates rather than known values, but expected production and sales costs may represent a fairly steady percentage of sales, at least within a reasonable sales range.

Especially in a first-pass model, most data cell entries may be proportionate or estimated values. As the project advances, data values in later models shift more and more to the determinate class.

With practice, you will find that modeling can be a potent tool. Manipulating a model teaches the dynamic way different factors of the model (and therefore the business) interact. You also can derive intuitive facts, for instance, that every time activity magnitude or duration increases, costs rise as well, but the relation can be far from direct. A good model accounts for items that do not scale perfectly when durations or contents change and for other real-world factors that distort simple pictures of cash flow. You certainly will build and rebuild your model many times. The first version usually predicts far too much cost compared to revenue, so the model must be adjusted in a way deemed helpful, and run again. A number of iterations later, a set of conditions (assumptions) should arise that begin to make sense, and subsequent runs will involve more refinement than gross adjustment.

One model that makes sense may not be sufficient to properly evaluate all factors with strategic company import. For the discussion at hand, we suggest without discussion the possible wisdom of creating at least one or perhaps several more plans, each of which makes sense. "Making sense" means that the basic functions of each seemingly sensible plan balance in a way that the sum of investment (early inflow) and revenue (later inflow) strike a fair balance with ongoing costs. Systematically comparing different models can provide powerful perspective and help in resolving otherwise ambiguous issues (e.g., determining the amount of required investment, balancing price versus sales volume, unit production costs versus production volume).

The current discussion centers on a very early time in a hopefully long-term project tenure. You have completed a product concept, and modeling is now being used to discover plausible businesses that could be built around the concept. Few facts are well defined, but it is shown later why a lack of particulars is of little hindrance. You may want to so some library research or interact with knowledgeable people to help better define model terms and thereby make the model stronger. However, the prime purpose here, very generally and admittedly with low precision, is to loosely characterize one or more businesses, defendable to investors, that could plausibly be built around the subject product.

We are making progress in the concept stage, having evolved from a primordial product idea through successive phases of enhanced product idea, idea feasibility testing, and creating a model or set thereof. The modeling, in essence, broadly expands the product concept so that it is just one part of a much richer and more complex business concept.

4.5 Marketing Research

In building a concept, all sorts of assumptions are introduced. Still, except for some early modeling, there has really been no hard evidence produced to support the real-world validity of a concept. For instance, it certainly would be helpful to know if the presumed customers, the source of all estimated model revenue, really would want such a product. Put differently, it is important to explore whether and to what extent the presumed market really exists.

Other important questions exist too, such as defining specific, customer-desired features, but those issues are addressed later. In the concept stage, your chief concern is to develop some comfort that a sound market really exists. The motive is self-preservation, because you are about to launch a major business development effort (prior effort is dwarfed by that to come). Proper preparation is mandatory, to protect investment in time, money, and emotional energy.

The central marketing research issue is verifying the existence of the envisioned market and sensing its size, questions that can be approached in several ways. Sources of useful information include trade magazines, technical journals, and casual or more formal interviews with key industry purchasers, executives, and sales representatives. In addition, trade associations can prove to be invaluable sources of information, as can exhibit functions at technical and trade meetings. Finally, direct or telephone interactions with potential product users can reveal insights that are otherwise inaccessible.

At this stage, marketing research usually is rather informal. However, highly proprietary product concepts enhance risks of inappropriate disclosure. While most people would not have a position or mindset to exploit your idea, some might. It is best to counter risk by asking people to sign a nondisclosure agreement (for details, see Section 6.6).

Always document all marketing research and its rationale at this or any other business development stage. For one, it may provide useful references of thought, actions, and results at later times. More to the point, comparing current and past records often can help you spot relationships and trends more easily and quickly. At any one time, you must collect enough information to draw dependable answers, but knowing when to stop is also important. At the concept stage, few past data may exist, so the end point basically occurs when market existence is verified, market size is roughly estimated, and a reasonable profile of various customer groups is known.

4.6 Detailed Evaluation of a Concept

You now have traversed the original idea, product concept and its feasibility, business concept, and, through marketing research, product and business concept verification.

You should at once be proud and humble. Pride is warranted for having persevered multiple project steps; it is difficult for the novice to fully appreciate the absolutely essential nature of every described task. Compliance demonstrates the discipline to follow conventional wisdom, without succumbing to seductive shortcuts. Nevertheless, the work is far from over, and a humble attitude is the best mindset. Concept evaluation should never end. As the company evolves, you must constantly challenge your personal and group reasoning and rationale. If a loophole in logic or action is uncovered, better sooner than later. More to the point, the recursive process of checking and rechecking pays strong dividends. Not only will such an approach enhance project perspective, it also will lead to the discovery of better ways to design, carry out, and manage a myriad of the large and small issues confronting every company function at all business development stages.

The discussion of conceptualization is largely finished. One last step is to integrate marketing research results into overall company planning.

The most concise documentation of company planning to date resides in the financial model. The model is a spreadsheet, a matrix of labeled columns and rows containing projected revenues and costs. At the intersection of a given column and row is an element called a *cell*. Each cell that is not a label or a graphic element contains a single number, such as a revenue item or a cost. The model does not include any technical sketches, drawings, parts lists, material specifications, supporting material for early cost estimates, or any other data form related to product development, design, production, or quality. Nor does it contain any marketing material, tentative sketches of promotional items, notes about pricing and other available estimates, or any other backup elements. Finally, the model also does not include any background administrative specifics, supporting material for the many rounds of planning completed to date. In raw form, the model core is just a rectangular array of cells, each containing a single number.

Inherent in the model numbers is a condensed version of all the other cited factors. For illustration, an estimate of R&D costs for a certain quarter depends on detailed planning for the whole company before, proximate to, and after the subject interval. R&D costs for the subject quarter cannot be modeled alone, nor can any other function in any quarter of the entire plan. It is not possible to build a reasonable model without thinking through every project aspect and how the various aspects interact, for all time periods of interest.

What is needed now is yet another version of the financial model, to fully incorporate all information and insight gained during the marketing research just completed and its interpretation. These marketing research results may affect many or all company functions, so the entire model should be analyzed and modified cell by cell, not just the marketing section, to accommodate all new information. Also, rethinking the model surely will lead to new thoughts that induce further adjustments, meaning that several new model iterations may be required to coherently meld all elements into a coherent picture. But that is to be expected, because any model rewrite rarely will be resolved in one pass. These efforts are not redundant or wasted. Every time you rethink any or all of the model, you build valuable perspective.

Thus ends our discussion of conceptualization and the concept stage of business development. In one view, we have merely described the creation and refinement of imprecise product and company notions, embellished with preliminary research and modeling, activities led by the administration division. We have not begun any detailed product design, let alone any preparation for production, quality, or other technology division functions. Nor have we carried out much detailed marketing division activities, presaging real-world marketing and sales.

On the other hand, the project has accomplished impressive advances. A raw idea, the primal flash was all there was at first. Now a sound, if preliminary, business concept is enveloping a feasible product concept. Both the product and the business concepts will mature further with company advancement, but an impelling level of substance and structure has been forged in the concept stage itself.

All the activities described in this chapter are the direct result of planning, led by administration's planning function and with all functions participating. A true planning milestone has been achieved, an early version of the company financial model having been structured, built, and completed. A later version of the same model forms a key part of the formal business plan, which is detailed in Chapter 6 and Appendix A. In the meantime, Chapter 5 builds on the introduction to planning presented in this chapter, by way of concept and concept stage review, to place this monumental topic in true formal perspective.

5

Strategic and Tactical Planning

5.1 Overview of Planning

Chapter 4 discussed the concept and its development, in essence the first step in planning, a topic this chapter addresses more formally. Recall that planning is one of the seven success factors (see Chapter 2). Heading the list of success factors is perspective, the most global factor and the one that broadly influences all the others. Planning is next, foremost among the remaining six success factors and exerting strong but more localized project influence.

In Chapter 3, planning was specified as all the activities used to define, forecast, and anticipate project detail and flow in terms of time and resource requirements. The treatment dissected that statement further to reveal more complete meaning. Planning was seen as acquiring external facts and combining them with internal facts and wisdom to estimate (model) the company's future. In other words, the known past and present, both outside and within the operation, are intellectually registered, analyzed, and interpreted to create a plan for future action.

There is no one right way to estimate the future or cope with it. Different individuals and management teams will decide differently and plot unique paths. But guidelines have arisen from conventional wisdom that can greatly aid the process. There are superior long-term preparation and problem-solving approaches that, while not telling you what to decide, offer tried and true methods of planning and decision making. For example, Sections 5.2 and 5.5 detail general approaches to management team building and strategic planning, respectively, but do not presume to dictate specific results.

The discussion of planning in Chapter 3 informally covered several topics, but it was intended merely as a brief debut of selected company issues and interrelations. We now concisely summarize diverse reasons for planning, setting the stage for later material:

- *Build the management team.* It is critical to progressively build the company management team so that both the present and the future are properly governed.

- *Set goals and priorities.* Most basically, planning must define all objectives and strategies and their relative importance for the company and each of its functions.

- *Compare past and future.* One seldom-stated reason for planning is to form a universal realm and language so that the past, present, and future can be realistically compared.

- *Forecast and meet human needs.* It is essential to gauge human needs at various points in the future and specifically plan how to meet those needs.

- *Forecast and meet resource needs.* Like human needs, future requirements of all physical resources must be gauged and plans made to meet those needs.

- *Model future operations.* Perhaps the most visible planning action is building and maintaining a detailed financial spreadsheet model of the business into the future.

- *Disseminate information.* The planning function must convey the results of its ongoing work to all key company people on a regular basis.

- *Measure performance.* Planning also must establish means to assess progress for a battery of defined measures, using business plan updates and other tools.

The rest of this section details each identified planning purpose, to clearly describe key issues and how they can best be managed.

5.1.1   Build the Management Team

An ongoing activity that typically begins soon after the primal flash is team building. In this age, group efforts are by far the norm; the time of single founders shepherding major ideas from advent to fruition is basically over. Still, every endeavor is different. Especially given the technical background of this book's audience, the effort possibly could be conveyed far down the road solely by the founder. Even so, operations of any reasonable size must eventually recruit some degree of permanent long-term help.

It is really the planning realm of a company's administrative division that builds the management team. Logically, the process should begin by the development of a plan for the future, formally recorded in part as a spreadsheet model (see Chapter 4). Then, the human resources needed for model fulfillment at each stage are extracted, the various functions are examined and grouped to reveal functional collectives potentially representing company departments or divisions, and qualified people to fill identified leadership roles are sought.

Usually, a fledgling enterprise shortly after its origin may have only the founder and perhaps one or two high-level, founder-recruited management team members, with no salaried employees at all. That situation may persist for quite some time, since early people tend to be ultimate management team members. They probably hold a company share and may be working with reduced or no pay. This form of operation is often essential, preserving precious cash for hard assets and product R&D rather than diverting it to salaries.

In any case, it is usually the founder alone who must find and recruit at least the first management team player. Those two individuals then build the management team further. For instance, a talented engineer may first attract a strong, upper management marketing person with an MBA, and then the pair act together to successively (over months or quarters) recruit a project head experienced in general management and finance, a separate expert in strategic management, and a top manufacturing executive, all with substantial experience related to the project's central product concept.

5.1.2 Set Goals and Priorities

A prime management team function is to set goals and priorities through the planning function. An early issue is to define the project and its prioritized objectives in writing. The point of departure is the product and business concepts, complete with feasibility analysis, marketing research, and preliminary company financial model discussed in Chapter 4.

It is critical that all planning activities and all other important actions of the company be written down. It is amazing how many people claim to precisely know their company's goals but cannot articulate them when pressed. Committing major and minor philosophies, objectives, policies, procedures, structures, designs, and many other operational elements to writing forces clear forethought and logic. The human mind has a startling ability to trick itself into cozy comfort with ideas and combinations of ideas that can be quite inconsistent. Transforming a thought to writing brings to bear heightened scrutiny that rarely gets applied otherwise.

With progressive company growth, planning must occur in an iterative process. Overall management team planning sets general direction, and then each department conducts more detailed planning focused on its sphere of responsibility. In turn, all department-level planning is reviewed at the management team level to carry the total company plan to greater depth. This global to departmental and back to global planning may undergo many cycles in any one major company planning effort, which usually spans weeks. In addition, prime planning activities should occur frequently (every several months) as the company moves forward, resulting in a planning profile with peaks at irregular intervals. Ideally, planning should never stop, and a good approach is for departments to carry on some meaningful level of planning all the time. Then, when major efforts are called out by administration's planning function, it is necessary only to ensure departmental plan coordination with company directives. In that way, overall planning is enhanced, as is perspective through continuous awareness and consideration of overall planning in the context of everyday complexity.

5.1.3 Compare Past and Future

Comparing items, events, and processes from disparate times is hard due to difficulties in rendering a fair comparison. A comparison of the design, production, or marketing efforts for two consecutive years may represent situations so unique that there are few if any common denominators to serve as comparative dimensions.

Much of planning's preparatory work may occur at the department or division level, carried out in the language and jargon of the division, and different divisions really do speak distinct languages. Within a given division or department, intercommunication for planning is much easier, since practically everyone knows the same terminology and has similar focus. That helps group leadership get everyone involved, regardless of job area or rank.

Alternatively, companywide, high-level planning led by the administrative division must use an unambiguous universal language transcending and comparably affecting all company functions. That means all actions must be reduced to a universal, easily measured unit of activity. The panoptic language of planning is money, permitting direct, realistic comparisons between the impact of diverse functions at different points in the past, present, and future.

We can now see a deeper significance to the financial model. The ultimate results of planning appear in three ways: (1) the financial model; (2) companywide strategic plans; and (3) policies and procedures (standard operating procedures, or SOPs) that help implement strategic plans by guiding and constraining activity. Even the strategic plans and SOPs, which are written documents, have a deep binding relation to money, being prime vehicles that control how the company builds and achieves value, cash flow, and short- and long-term viability.

5.1.4　Forecast and Meet Human Needs

The planning function uses past and present constructs, combined with interpreted research and insight in light of changing internal and external forces, to forecast the future.

One premier planning component is predicting human needs, given a parallel plan that specifies tools, equipment, and other physical supporting elements. Past and present experience may help, but you also must account for changes in productivity due to advancing equipment, procedures, employment laws and agreements, makeup of the work force, and other relevant results of technological progress. Finally, all data must be combined and interpreted to plot a time-based estimate of future human needs.

You cannot estimate future company performance without fairly specific projections concerning every company component. Total company performance is critical for various reasons. In many cases, proper supplier agreements must be in place long before actual needs, and the same is true for capital arrangements, labor force needs, legal considerations, facilities requirements, and much more. This is yet another compelling reminder of the criticality of planning.

Once planning has plotted future human needs, a plan must be developed to secure them. Today, you cannot always hire the desired numbers and types of employees arbitrarily. Sometimes the labor market is extremely tight, as it is now for, say, computer programmers. If any tight future hiring windows for some specialties are anticipated, appropriate responses must be designed and implemented early to orchestrate solutions. If hiring is difficult, potential recruiting aids include higher salaries or benefits, flexible or other forms of attractive hours, provision of day care and other assistance for working parents, allowing worksite naps during personal time on the job (lunch, breaks), and so forth. All these factors carry some kind of cost, which must be factored into the master financial model at the appropriate time and level.

5.1.5　Forecast and Meet Resource Needs

Human needs usually are just part of total company requirements. Nonhuman resources can include physical facilities, major and minor equipment, office furniture, computer and communications systems, transportation equipment, safety systems, and perhaps much more, depending on the business.

Analogous to human requirements, the past and present are often useful in estimating physical resource needs to meet certain activities and throughput levels. Even so, management must stay abreast of emerging resource improvements, new techniques, and broader factors such as raw material and component availability, pending legislative actions, competitive improvements in productivity, transportation issues, and any other relevant elements. Management additionally must combine the diverse data to plot a time-based estimate of future physical resource requirements.

Changes in physical resources can be tracked several ways. Trade associations are an excellent source, so join one or more trade associations applicable to your operations as soon as resources permit. Also, trade and technical magazines and meetings in your field are paramount for monitoring change. Finally, university researchers or groups may be particularly interested in your realm, free of proprietary bias. Make friends with them and stay tuned to their advancing work. With luck, this help can be free or at least reasonable. Universities may take on parts of your research for expert-supervised student projects, addressing physical resource problems (and even contributing to human resource needs as well, through student projects or faculty consultation). Sometimes proximity to a university with strong potential interactions can form a compelling location parameter.

Planning must first estimate the physical resource requirements mapped over future time, decide how to secure them, and see that they are in place and functional when needed. Long lead times often are involved, reinforcing the criticality of employing professional planning methods. Facilities, equipment, laws, regulations, vendors, customers, and a host of other key factors can change wildly. Change is inevitable and can work to the advantage of those properly prepared. Also, monitor your competitors closely; change may catch them off guard, providing you with golden opportunities to gain ground almost for free.

Planning predicts future physical resource needs by using the many reconnaissance methods described here, such as trade associations, publications, and academic experts. Professional experts (paid consultants) can be useful, too, but try other more economical means first. Data from the described methods must then be interpreted, reducing it to focused conclusions. Subsequently, model the one or perhaps several scenarios that management deems worthy of detailed pursuit, tuning the model or model set to study how different business assumptions or approaches play out. Finally, when one or two models begin to show promise, devise strategies to test competing models for growth, robustness in the face of problems, abilities to face various competitive moves such as price reductions, and so forth. For instance, competing models can be tested by equating sales and comparing variations in selected costs or equating key costs and comparing profit. Also specifically evaluate the various assumptions of one or multiple key models for validity. One way is to systematically vary assumptions (e.g., sales, COGS, wages and salaries) to note the effect on financial performance. Another is to use competitor product literature, trade association data, government demographic and other data, reverse engineering of competitive products, and so on.

5.1.6 Model Future Operations

Modeling has been described as the most visible planning activity, yet it is only one of the eight prime planning topics under discussion. Modeling is prominent, but it merely allows summarizing planning results clearly and concisely. An introductory orientation to modeling was presented in Chapter 4, and much added detail about financial spreadsheet structure and content will be provided in Section 5.3. Here we comment on how modeling can be effectively conducted.

We live in the remarkable information age, the foundation of which is the computer. Given the ubiquitous nature of personal and professional computers, it is virtually mandatory that all company financial modeling be computer based.

In modeling, significant time is required to design the structure and build detail of the initial spreadsheet table. This involves forming a new spreadsheet, creating labels for all columns (time intervals) and rows (revenue or expense line items), and entering an initial set of values in the table defined by the labeled columns and rows. Once the model's basic architecture and general content have been established, changes of detail can be entered and results generated much more quickly, making it practical to produce frequent model updates and to conduct easy what-if trials that compare different approaches.

Accepting the use of the computer in general, we also suggest using an IBM-compatible system or at least one that can emulate the DOS or Windows operating systems. While alternative systems can produce compatible results, it may be useful at times to pass model file copies to other people, such as accountants, attorneys, bankers, and investors, who seem to strongly favor DOS or Windows-based operating systems. Also, much more accessory software is available for IBM-compatible computer systems than for other computer architectures.

5.1.7 Disseminate Information

Every company functional group participates in certain planning activities from time to time. However, administration's planning function is global, affecting the whole company. Much work occurs there that is never seen elsewhere. So administration's planning function must disseminate key information to all other company groups.

Probably the most obvious element produced and distributed by the administrative planning group is the business plan. Most people think the business plan exists mainly for external reasons such as funding and orienting key outside professionals. The business plan serves many internal functions, too, but it just is not formatted properly, presented in sufficient detail, or updated often enough to serve as an internal planning news media. Planning should supply ongoing summaries and interpretations of marketing, sales, technical, administrative, cash flow, and many other issues to all concerned. Some news should be broadcast to every company function, while other data must be distributed selectively. It is important that everyone know about major events or tactical programs that are ongoing or planned, for proper companywide functional preparation and support. However, certain information, such as critical future projections or strategic plans, may dictate a restricted audience.

Mature companies usually include a formal accounting group or at least a function that organizes raw data for an outside accounting and perhaps payroll management firm. We recommend that accounting and payroll activities be placed within the planning group. That facilitates budgeting activities, originally designed by planning but managed and reported by accounting. It also facilitates integrating ongoing planning into updated budgets and tracking planned versus actual financial results.

5.1.8   Measure Performance

Another prime planning function is to track company progress over time. Progress measures can target overall performance or more specific actions, as long as they are quantifiable. Qualitative comparisons can be used to elaborate quantifiable measures, but they are more difficult to resolve and interpret.

One major measure of global progress is to systematically compare the periodic financial records that the company already creates for tax and investor reporting. Another measure of global progress is comparing historical and current versions of the business plan. In that regard, it is helpful to structure the business plan's financial model to correlate well with the accounting and reporting methods employed to report company performance. Finally, a general productivity index, such as annual revenue per employee, can prove useful in assessing operations efficiency.

Turning to specific measures of progress, one example is manufacturing productivity, a measure of throughput versus total production cost tailored to your specific situation. A second index is total product design costs versus annual and lifetime sales for the same product, or alternatively a record of ongoing costs of quality (see Section 13.5). In the marketing sphere, comparing total marketing costs and/or direct cost of sales versus sales revenue can be instructive indices of efficiency, especially if estimates of competitor efficiencies can be generated. These measures allow management to check the plan against actual results and to adjust methods when problems become apparent. Finally, one good administrative indicator of progress and performance would be tracking QA, such as auditing compliance of actual company practices against written SOPs governing all company functions.

The monitoring and control of various company functions are addressed comprehensively in Parts III and IV. There, formal systems of control are described in detail, and general implementation methods are developed. However, during early business development, less formal controlling mechanisms designed along present guidelines should be installed and directed by the management team. We have described approach philosophy and some general examples as a guide, but situation diversity dictates that you rethink and redesign unique measures for a given real-world endeavor.

5.2 Building the Management Team

By way of background, one crucial planning element is recognizing the founder's personal versus business goals. Similar considerations apply, sometimes more sparingly, to major players who join the management team, especially the earliest entrants. Compared to business goals, personal objectives often get slighted, since there is usually an early rush to start the project. That leads to various plans and accomplishments, focusing generally on, successively, the primal flash, the product concept, the project concept, and the business concept. As the succession and the concept stage advance, the project progressively assumes a life of its own, leaving personal considerations wanting. Individuals may have varying family, social, cultural, recreational, religious, and other interests and commitments, along with different lifestyles and short- and long-range objectives and strategies, diverse factors that can affect company performance and dedication.

It is incumbent on first the founder and subsequently the evolving management team to properly balance personal and business goals. Personal elements vary widely across projects, but key issues to consider include company location, work hours for various people, salary versus ownership, and short- and long-term personnel benefits. Other elements for attention might be part- versus full-time effort, personal contributions of nonpersonal resources, telecommuting, and rewards for outstanding performance. The list could be greatly expanded. The point is to be sensitive to the special needs and desires of key players, not just the business. It is not necessary to honor every whim, but careful attention to select wishes and needs can build tremendous respect, loyalty, and handsome paybacks.

Attention to personal concerns can aid management team building substantially. At first, the founder may be completely alone. Recruiting the first team player beyond the founder can be the most difficult, since there often are less-tangible project artifacts or ideas to show. Also, there is likely little or no funding in place, so overall, the project is largely mental rather than tangible.

Recruiting at this stage, therefore, requires relatively large rewards and perks, to counter the high risk and uncertainty of early involvement.

One way to overcome several hurdles at once is allowing part-time or off-hour involvement by one or more key players. That way, recruits can protect themselves with ongoing income, while building an opportunity for large future returns. Another way is to secure strategic partners, individuals, or firms willing to devote part-time help for an ownership share or other potential future benefits. As shown in the war story in Appendix B, acquiring strategic partners and using part-time management team members are potent ways to control costs while making substantial early progress.

Team building is a slippery road, but several well-tested approaches have proved effective. The secret is early detailed planning on your own to define what is needed most and then recruiting capable individuals to join in making it happen. Remember, not only must they be experienced and skilled, they must have the credentials that make their ability obvious to outsiders. Next, devise a necessarily unique strategy to make your specific project attractive. Be completely prepared, organizing existing materials, research, design sketches, market research, or anything else that will convey as detailed a picture of the past, present, and future project as possible. Then and only then, when you are fully prepared, make your pitch. Carefully pick the time and the place to optimize your image and the project's image and to ensure the prospect's full attention. If you do not succeed with your first choice, successively move down a prioritized list, expanding it if necessary to maintain the search. Persistence is the watchword. Refine the presentation, revise the plan, extend the list of candidates, and above all keep trying. Success will come eventually if you have truly developed a sound and logical project concept that meets all seven success factors described in Chapter 2.

Besides funding, team building is typically the hardest project task, so kill two birds with one stone. Funding difficulty is substantially reduced by steadfastly accepting no less than a sterling management team. Recruit a stunning team meeting the criteria defined in Chapter 13 that also has significant access to potential investors. By choosing the team wisely, you can preempt many, if not all, funding hurdles—a powerful management team can impressively command requisite attention and deliver a compelling investment package.

After the founder secures the first additional team member, further recruitment becomes slightly easier because the project is stronger. Each later addition makes the team and the project stronger and more attractive to recruits. However, that works only if every member meets the demanding criteria of top experience, skill, credentials, and, if at all possible, access to potential investors. Note that the opposite is true as well: Even one weak member defeats the rationale and renders the project and the process limp. Accept weak team members early, and you probably will not be able even to match their limited caliber later or find any investors. Team building is a decidedly nonlinear function that markedly accentuates early results.

5.3 The Model and Its Structure

The basic financial model structure was sketched in Chapter 4. This section describes the generic model structure in more detail.

The generic spreadsheet in Figure 4.2 is a standard financial model (standard model for short). Any analogous spreadsheet possesses a core area featuring numeric data entries. Figure 5.1 represents a nonspecific data-area patch in such a standard model. As intended for all models and emphasized in Figure 5.1, the standard model's core area features a rectangular array of rows and columns. The rectangular area where a specific column and a specific row intersect is a spreadsheet cell, or just cell. The content of core area data cells is defined by labels positioned across the top (to identify columns) and down the left side (to identify rows) of the spreadsheet. (Assume that the standard model is created by an IBM-compatible spreadsheet software package.)

All popular spreadsheet computer programs operate similarly. Any single cell can contain one of three types of entry, depending on program viewing mode (Figure 5.2). One is an alphanumeric string (letters and/or numbers) interpreted as text, usually serving as a column or row label but other uses are possible. Second, the cell may feature numeric data, the type of data and its units being specified in the column or row label. Third, the cell may display a formula.

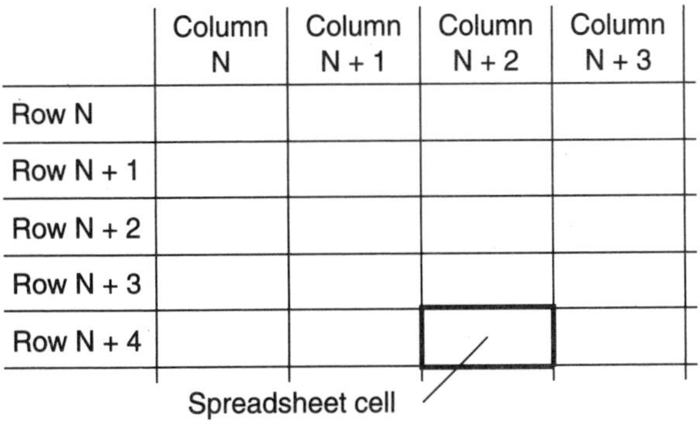

	Column N	Column N + 1	Column N + 2	Column N + 3
Row N				
Row N + 1				
Row N + 2				
Row N + 3				
Row N + 4				

Spreadsheet cell

Figure 5.1 Numeric entry (core data) area of a standard financial model. Each column is uniquely defined by a column heading above (not shown) and a row head to the far left (not shown). The singular data entry field at a row-column intersection is called a spreadsheet cell, or cell.

A Alphanumeric entry

Profit before tax

Any string of letters, numbers, and other symbols
or spaces in any order. The string is interpreted as text, the illustration
showing the cell used to define the "Profit before tax" row heading for the
standard financial model in Figure 4.2. Alphanumerics are left justified.

B Numeric entry

123.45

The core area of the standard financial model in
Figure 4.2 is largely or entirely composed of numeric entries. Financial
models primarily involve numeric entries representing dollar amounts (e.g.,
$123.45), right justified to distinguish them from alphanumeric entries.

C Formula entry

12345.67

A formula entry is signified by an entry-leading
algebraic sign, followed by the formula proper. A formula entry leads to two
displays, the formula in a heading area of the spreadsheet display and the
executed formula result in the subject spreadsheet cell. The example shows
a "Profit after tax" result, computed from, say, the formula E12+E18+E27+E38
(the sum of numeric results displayed in rows 12, 18, 27, and 38 of column E).
Consistent with the example, the latter rows are intended to represent totals
for each major spreadsheet group (Revenue, COGS, Direct sales costs, and
G&A expenses).

Figure 5.2 A spreadsheet cell can accept one of three types of data entry: an alphanumeric
string, numeric data, or a formula. (Lines and other appearance elements are alphanumeric entries.)

In fact, the formula displayed in the spreadsheet program's function viewing
mode is precisely that used to calculate the same-cell numeric entry displayed in
the data viewing mode. A cell displaying an alphanumeric entry has no associated
formula, whereas numeric data can be either direct entry of a number (if there is
no underlying formula) or a calculated value (if an underlying formula exists).

A standard model is formed by defining all spreadsheet cells for a given situation consistent with Figure 5.2. Because you create the actual structure (general layout and specifics) of the entire spreadsheet, the content is completely
controlled. The spreadsheet can be expanded or contracted at any time, by
adding or deleting rows and/or columns. First, reserve several entire rows at the
very top of the spreadsheet for a title and any qualifiers (e.g., revision number,
date, company division or segment, set of assumptions). Then build row and
column headings appropriate for your endeavor (see Figure 4.2). As shown, the
column headings relate to time and usually represent either a single time period
(e.g., a quarter) or a sum of such entries (e.g., year 1). Row headings vertically
arrayed at the figure's far left are more diverse, but need not vary much from the
example unless desired for special effect. To emphasize key relationships in our

standard model, individual row headings have been collected into several groups, including total revenue, direct sales, cost of goods sold, and so forth (see Table A.1).

In Figure 4.2, every indication of a row or columnar data total requires building a formula into the cell to carry out the intended arithmetic operation. Many other cells may have formulas too. For instance, you may decide that every entry across a row labeled "Legal fees" (and therefore across all spreadsheet time intervals) are to equal 1% of total revenue for that same period. You would then enter the appropriate formula to calculate that value in each (nontotaling) cell of the row whose label is "Legal fees." You build the model by first defining every column and row with appropriate headings, adding lines and other graphic elements for visual effect, and then entering a numeric value or a formula in every cell assigned to hold data. Manuals for popular software packages provide comprehensive coverage of spreadsheet construction.

A particular model is characterized by its basic structure and by a set of specific assumptions. You can set up any number and complexity of assumptions, rows and columns, and horizontal and vertical subtotals and totals that you wish. For clarity, this discussion adheres to the specific structure depicted in Figure 4.2.

By adopting the layout in Figure 4.2, we have fixed model structure, being rigidly defined by the column and row heading definitions shown in the figure. Even so, an infinite number of distinct models is still possible, using the singular Figure 4.2 structure but employing a different set of assumptions in each case.

An assumption is any number or formula entered into a core area model cell that does not represent a known fact. We may know the costs, for instance, of facility, legal, and accounting expenses, from experience or as the result of research. Still, we do not know exact unit sales in any one time period or the direct costs to sell those units. The first group of factors, facility, legal, and accounting expenses, are not assumptions, the costs being accurately known. However, unit sales and direct selling costs for any time period are uncertain, educated guesses, qualifying as assumptions.

Actually, a given model usually involves many more assumptions than our illustration. A single model of the complexity discussed in Appendix A might contain 50 or more assumptions. The permutations of possible assumption sets are basically intractable (50 items taken 50 at a time, with say 50 variations of each item). For practicality, choose one or a small number of assumption sets with which to work. Use of one assumption set results in one model. In turn, to compare two or more different situations, build several models involving an identical structure but employing some systematic variation of key assumptions.

The standard model being defined here involves three financial statements. The world of finance recognizes a number of standard reports or data structures that are called *statements*, three being universally recognized in business development:

- *Income statement*, a tabular summary of all revenue and expenses, by time period (months or quarters);
- *Balance sheet*, a tabular summary that equates total company assets to the sum of liabilities and equity for each time period depicted in the income statement;
- *Cash flow statement*, a depiction of the actual status and movement of cash for each time period displayed in the income statement.

In the privately funded arena of business development, statements are designed to cover either 5 or 3 years, 5 years being mandatory for all but the smallest projects. (This book universally assumes 5-year statements.)

The standard model in this book is based on *cash accounting*. Another popular form of business accounting is *accrual accounting*. The distinction between the two is important so you can correlate the information in this book with other writings.

Cash accounting counts revenue when it actually arrives and expenses when the bill is actually paid. This is straightforward, the way you likely keep your personal records. Thus, the company's main checking account journal agrees exactly with an income statement prepared to end on the same date. The balance sheet maintains a running history of assets, the total being equated to the sum of liabilities and equity at the same instant. The cash flow statement is, in essence, a record of cash movements duplicating certain income statement entries but organized to strictly and meaningfully depict cash flow and avoid some noncash elements of many income statements.

In contrast, accrual accounting records expected rather than actual revenue and anticipated rather than actual expenses. For instance, the company records an expense when the obligation is incurred, not later when the bill is paid. In accrual accounting, relevant income statement entries do not reflect actual cash movement. As another example, revenue is counted when a salesperson books a sale, not later when the money is received. The cash flow statement is particularly critical under accrual accounting, to reconcile the income statement to reflect perhaps dramatic discrepancies regarding the true flow and status of cash.

Accrual accounting has operations and tax advantages once the company has grown significantly. We suggest beginning the operation under cash accounting, switching to accrual accounting later if professional accounting advice so dictates.

5.4 Estimated Funding Requirements

One purpose of the company's financial model is to estimate funding requirements, by simulating every function throughout the 5 years of the standard model.

As described in Section 5.3, the model projects both revenues and costs. Early in the project, probably spanning many quarters, there will be no revenue from sales at all. Even when sales revenue finally begins, the amount may be far less than total costs, a situation termed *negative cash flow*.

Because negative cash flow will not work, you must secure sufficient money independent of sales revenue, until sales revenues completely support operations. There are two prime nonsales revenue sources. First, self-investment by management team members is theoretically possible but seldom practical. Any such internal investment should be taken first, if possible, and duly rewarded with stock. The second investment source is outside funding, requiring clear explanation and rationale for exact money requirements, when needed, and its intended use. The written business plan provides that information, supplemented with verbal discourse, formal presentations, and other company hard-copy material.

This exemplifies again the critical nature of ongoing, in-depth planning and maintaining a current written business plan. Planning is needed for the management team proper, to provide a framework for action and decisions. It is also indispensable to lure qualified investors to the project. The business plan is the most concise and structured representation of the planning process and speaks loudly to the competence and organization of management itself. A stunning plan creates a compelling image of the management team and the whole project. A weak or poorly written plan instills negativism about the entire endeavor that probably cannot be remedied by any amount of courting or additional material.

5.5 Strategic and Tactical Planning

Strategic planning is the defining and implementing of long-term company objectives by the administrative division's planning function. Business decision making is complex, involving in part a constantly changing multitude of the following factors:

- *Internal factors.* These factors are the collective array of all functions that make up the company proper, dispersed in our model through the company's three functional divisions (administrative, technology, marketing).

- *Immediate external factors.* External but nearby influences include competitors, suppliers, government and other regulatory agencies, advocacy and consumer groups, and customers.
- *Remote external factors.* These factors are more distant than immediate external factors but still of company influence, for example, social pressures, economic conditions, political influences, and technological developments.
- *Stakeholders.* Stakeholders have distinct influences, often with objectives mutually at odds, for example, stockholders, owners, management team members, employees, strategic partners, communities, and the larger populace.

All those factors must be studied, monitored, evaluated, and melded into the decisions and actions of top-level management. The potential exists for considerable uncertainty about the best short- and long-term direction.

Consider this real-world illustration. A founder-as-god (see Chapter 1) suddenly decides to open a second operation 200 miles from a young existing facility. The action is not questioned because the founder-as-god has picked a weak management team and structured the board of directors with the same people. The decision ignores several stark realities, including lack of resources and staff (internal factors), improper demographics at the new location (immediate external factors), a laggard economy in the new location and difficult interfacility logistics (remote external factors), and the strategic threat to viability of further spreading an already thin operation (stakeholder interests). Company bankruptcy is the ultimate result.

Sound business decisions are not made in isolation. Through C^3 (coordination, command, and control), top executives of large, mature companies employ a formal process termed *strategic management* to intentionally develop and continually update data and interpretations about the company's universe. *Strategic planning* is a recognized subfunction of strategic management directed to devising long-term (many years) and short-term (annual) general direction. Strategic planning creates companywide plans deemed best to achieve company objectives, permitting improved anticipation and superior future readiness. Functional managers then plan more localized strategies and tactics for implementing companywide strategic plans and governing their implementation. As an analogy, strategic management (the pilot and copilot) command the company (fly the airliner) according to the course (flight plan) plotted by strategic planning (the navigator). Strategic planning also provides continuous feedback (e.g., navigation data, weather updates, emergency information) to strategic management (pilots), so they can effect any required midcourse corrections (altered flight plans).

This book has adopted elements from the excellent reference *Formulation and Implementation of Competitive Strategy*, by J. A. Pierce II and R. B. Robinson, Jr. Strategic management and strategic planning are far too complex for adequate coverage here, but they are central themes of Part IV (Chapters 12–14) for optimizing control and building company value. Note that what we call C^3 is essentially congruent with strategic management and strategic planning taken together, formal functions usually reserved for association with more mature companies beyond the startup stage.

Strategic management and planning attend to the following nine essential factors:

- *Company mission*, a written statement that sets out broad declarations of company purpose, philosophy, and goals;
- *Company profile*, a written description that defines and sets boundaries on the internal conditions and capabilities of the company;
- *External environment*, a written description and interpretation of the company's external environment, in view of both competitive and general contextual features;
- *Preliminary options*, an analysis of the possible options suggested by comparing the company profile to the external environment;
- *Final options*, an extension of the preliminary options, to define the resulting options when the company mission is folded into the critique;
- *Long-term objectives and strategies*, a defined set of long-term objectives and grand strategies required to achieve the final options;
- *Short-term objectives and strategies*, annual objectives and short-term strategies compatible with long-term objectives and grand strategies;
- *Implementation*, imposition of resource limitations to define ways of implementing the objectives and strategies, by judicious matching of people, tasks, structures, technologies, and reward systems;
- *Evaluation and control*, review and assessment of the overall strategic process for effectiveness on an ongoing basis, as a cornerstone for control, and as a critical input for future decision making.

Strategic management and planning devise, design, direct, organize, and control companywide decisions and actions. The result is the company game plan and a solid framework for managerial decisions.

It is important to contrast the strategic plan and the business plan. Strategic planning is detailed, documented companywide preparation that draws on all relevant external and internal influences to determine general direction, broad goals, and expected accomplishments for every company function. A formal strategic planning function usually is not formed until late in the startup business development process, but the same fundamental planning principles are integral to the business development recipe presented here in a less formal way. The business plan (the prime topic of Chapter 6) is typically a written, bound treatment extracted from the (formal or informal) strategic plan, addressing a narrower and less detailed company view targeted to employee orientation, funding, or other defined purposes.

5.6 Formal Business Organization

The primal flash evolves first into a product concept, and then progressively matures to take on both product and business concept aspects. A business concept must emerge too, because projects must gain significant size to have a realistic chance of commercialization. You can try remaining a one-person shop by simply subcontracting virtually all activities except perhaps for ultimate coordination and control, but it is the rare project and founder that are suited to such an approach.

We reiterate that almost all product-oriented businesses assume the formal organization of a corporation, but other structures are possible. We discuss several alternative product-oriented business organizations, along with the pros and cons for each (e.g., see these references listed in the Suggested Reading section: Baird, *Engineering Your Start-Up*; Vesper, *New Venture Strategies*). Only book-relevant issues are treated, so it is important to consult legal or accounting counsel for advice regarding your specific case.

5.6.1 Sole Proprietorship

A sole proprietorship is an entity owned and run by one person, but employees are permitted. Basically, the owner is the business. This is the least expensive business type to form and operate, since only appropriate licenses are required. A sole proprietorship is not taxed separately; the operations are reported on Schedule C filed with the owner's tax return. Roughly 70% of U.S. businesses are sole proprietorships. The owner has full liability exposure that extends to personal assets.

5.6.2 Partnership

A partnership is owned by two or more people, and additional employees are permitted. Each partner files the proportionate share of operating costs and revenue on his or her personal tax return; under proper circumstances, the distinct partners can divide income and expenses in a different ratio than percentage ownership. Written partnership agreements are essential for mutual protection. About 10% of U.S. businesses are partnerships. Each partner has liability exposure that can extend to the entire operation and even beyond. Special forms of partnership exist in some jurisdictions (e.g., limited partnerships) where unique liability and other legal elements prevail.

5.6.3 Limited Liability Company

The limited liability company (LLC) is a reasonably new type of business structure that is now valid in nearly every state. The LLC is in essence a cross between a corporation and a partnership. The LLC enjoys limited liability advantages of corporations (see below) but features the flexibility and tax reporting advantages of partnerships. The LLC is very popular for certain small businesses, but it is not favored when large amounts of capital must be acquired from many distinct sources.

5.6.4 Corporation

A corporation is a legal entity that exists separately from the people or other legal entities that own it. The owners (shareholders of corporation stock) generally incur no personal liability for debts or lawsuits against the corporation, limiting shareholder risk to their stock investment. The ability to issue common stock and other types of securities provides many more funding and growth avenues than other organizational types. About 20% of American businesses are corporations.

A corporation is a distinct legal entity and is taxed as such, so taxation occurs twice, once for the corporation itself, and once to shareholders that receive corporate dividends. All or some of the double taxation can be overcome for stockholders who are also employees, but double taxation is usually still significant for profitable operations.

Technically, there are two classes of corporation, the C-type (C Corp) and the S-type (S Corp). The C Corp is basically as described above, and the S Corp is a special form that receives most of the advantages of the C Corp. However, by meeting certain requirements amenable in many startups, the S Corp avoids double taxation. S Corp owners are allowed to report their share of revenue and expenses on personal tax returns, in a manner analogous to a sole proprietorship.

The majority of projects assume the corporate form of business organization to avail themselves of flexible funding opportunities and limited owner liability. To preserve the early theme and avoid undue descriptive complication, this book continues to assume a corporate form of business organization.

6

The Business Plan

6.1 The Rationale for and Nature of the Business Plan

Leveraging the in-depth presentation of planning in Chapter 5, we now focus on the business plan. This is a formal written document that summarizes a selected subset of historical and planned activities at a particular time and for a defined audience.

It is instructive to carefully dissect that definition. The business plan is a formal written document that is a recognized entity in the business development community. Relevant people understand reference to a business plan, picturing a written compendium that describes an intended or existing business, how it is structured, what the business does, and the company's history, current status, and planned future activities. Also presumed is that the business plan contains a financial summary of past activities and projections that estimate the costs, revenue, and cash flow for time periods extending well into the future. In addition, it is expected that the business plan references a specific date or time period (a specific month or perhaps quarter or season), because the company and its operations change with time.

Business plans can be found in various lengths and complexities. Some are simple and just a few pages long, a makeup unsuited to the privately funded tier of business development. At the other extreme, some business plans in the venture capital sphere comprise several phone book–size volumes, excessive for our purposes. The typical privately funded project requires initial and later business plans that most likely occupy one bound volume, 30 to 75 pages in length, including appendixes (Figure 6.1).

Discussion in previous chapters has loosely addressed the need for a business plan. A detailed discourse on the various purposes of a business plan is now

(a) External view of bound business plan.

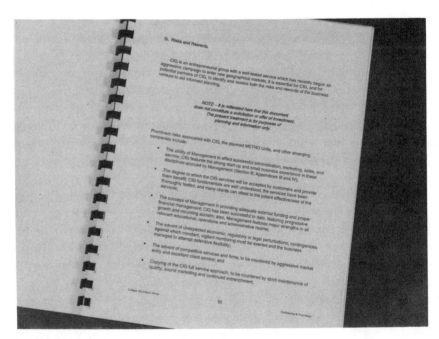

(b) Narrative page showing style variation to break up text.

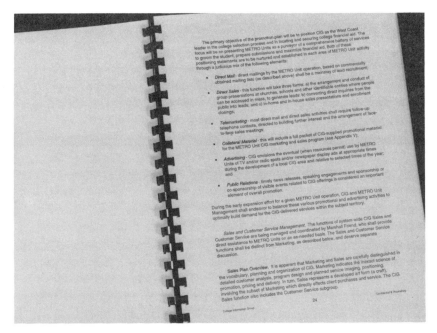

(c) More narrative; bulleted list for variation.

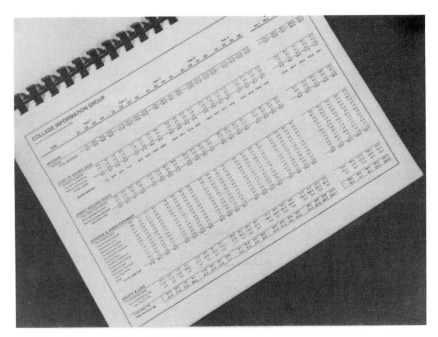

(d) Sample page from financial model.

Figure 6.1 Photographs of a business plan and key plan sections.

in order, to serve as a central reference and educational tool. Those purposes are as follows:

- *Funding.* The business plan forms the centerpiece of project or company documentation used for the acquisition of funding.
- *Frame of reference.* All decisions and actions by management should strictly adhere to business plan dictums.
- *Out-of-house communications.* The business plan becomes the primary company-detailed descriptive tool for dealing with vendors, key customers, and business professionals outside the company.
- *In-house education.* It is crucial to educate many nonmanagerial people in the company regarding future direction, and the business plan is the best place to start.
- *Standards of performance.* Measures of company and function progress with time can most easily use the business plan as a point of departure.
- *Means to track planning.* A quick overview of planning effectiveness at various historical times can best be done using past business plan versions.
- *Departmental plans.* The business plan for the whole company forms the master gear against which all departmental plans must precisely mesh.

The business plan exerts broad and deep influence on diverse company functions. Each cited consideration is expanded in the following subsections.

6.1.1 Funding

The topic of funding enjoys broad attention in this book. The goal here is to develop ties between the formal written business plan and the constellation of activities assigned to the realm of funding.

In dealings with potential investors, it is obligatory to communicate a tremendous spectrum of both large and small company details. A disorganized presentation causes two problems. First, it is difficult for the data recipients to assimilate the information, since they must not only struggle to capture each fact but also attempt to reorganize the material on the fly. Second, disorderly information reflects badly on the company, the product, and management. A poor presentation makes it difficult for recipients to tell if the project is good or bad. In any case, the project usually is disregarded without further review.

There are many facets to interacting with potential investors. The total collective of communications channels may include elements such as equipment or process literature, prototypes or mockups, tours, demonstrations, reviews or demonstrations of competitive systems, visual aids, discussion sessions, and written documentation. To ensure an orderly and logical presentation, it is best if the recipient first reads the current version of the written business plan, to provide a scaffolding for the data. The business plan is the best framework available, covering all company aspects in sensible, concise, and complete form, thereby providing excellent orientation for remaining aspects of company presentation.

6.1.2 Frame of Reference

The critical importance of management consistency has been emphasized before, but the paramount need for statement and action uniformity is easy to overlook. Everyone on the management team must be on the same page; even small discrepancies expressed by various team members will be glaringly apparent to investors.

Regarding investors, disharmony can imply a number of underlying troubles, none of which is acceptable. Examples include disagreements of direction, personality conflicts, poor planning, ineffective communications, and lack of understanding of plan import. Be mindful of two points. First, do not have any disharmony. Second, when you have disharmony anyway, solve the issues behind closed doors.

Private disagreements probably are inevitable and can be quite healthy if handled properly. A management team of founder clones or lackeys under the thumb of the founder is not a true management team. A management team must be multidimensional and forceful to bring maximum analytical prowess to bear on widely diversified subjects.

The business plan serves as a central frame of reference for the entire management team. The business plan must be real, not some trumped-up smoke screen designed purely to lure money. That means the management team must religiously use the business plan to guide the incessant daily barrage of required determinations. Decisions, large and small, are what management is about, and the business plan sets moderate and often tight constraints on available courses of action. In a sense, the business plan is a set of templates. If some intended action does not fit one or more defined patterns, it is either unacceptable or must be raised with general management as a possible basis for plan modification.

6.1.3 Out-Of-House Communications

The company will have significant and recurring need to interact in one way or another with numerous outside entities. Relevant people or groups may include commercial and investment bankers, accounting and legal professionals, advertising and marketing entities, vendors of technical products, materials and services, potential investors, prospective high-level employees, and consultants.

Some means must be employed to orient people quickly and thoroughly on the operation. The written business plan is by far the best vehicle. However, in many cases, the business plan may provide too much information. Often the outside entity needs to know only select information within limited company realms. For such purposes, perhaps excerpts from restricted areas of the business plan will do (e.g., a briefing package, road show, or selective breakdowns of each division's activities). In other instances, the business plan may not suffice at all, other means being required to convey requisite data.

Be wary. Business plan material is highly proprietary and confidential. Section 6.5 discusses protection for the company and its intellectual property, a discourse highly applicable to business plan content.

6.1.4   In-House Education

As the company grows, the count of nonmanagerial people increases much faster than the roster of managers. Many a nonmanagerial person functions where knowledge of company planning and direction would be of great benefit. Lead engineers, marketing managers, and top accounting people are possible examples. The business plan may represent an excellent introductory tool to begin bringing such people current. Still, the business plan is highly proprietary, so exercise careful thought in deciding who sees what.

6.1.5   Standards of Performance

Projects with some history may have available past business plan versions that contain many plans, predictions, and expectations.

In one view, the business plan can be said to define achievement goals rather than achievement forecasts. In fact, we argue throughout the book that the business plan must be taken seriously, so it is a trivial step to carry the latter thought to the limit. In that regard, we take the business plan to be the expected result, in essence, the drop-dead target of project or company performance. Furthermore, that is precisely the way most investors expect the business plan to be treated.

Planning is of little use if it is not meant to derive and depict realistic performance targets. Most circulating business plans are grossly optimistic. Professionals who regularly review plans often divide revenue figures by 2 or 3 as a rule of thumb, and they also know that the costs and development times for various activities usually are underestimated substantially. Experience teaches that most plans lack the insight of extensive startup experience. Imagine the impression you can make if you craft a business plan that actually passes muster. Suddenly, you enter an exalted realm indeed.

6.1.6 Means to Track Planning

Since we are taking the business plan to be the cardinal company road map, any past plan versions can be very useful for comparing predictions to actual results. First, a past version provides a means to qualitatively and quantitatively track business plan adherence. Second, deviations from predictions can be used to focus attention on discrepant activities. Finally, the process provides a powerful way to assess planning methods.

Tracking business plan projections versus actual results is an invaluable and virtually mandatory activity. One most wants to follow the efficacy of total planning, but the concise form of the business plan material may make its review the best way to at least begin the process. Nevertheless, many planning elements are too detailed for business plan entry but critical to tracking planning performance. Management should decide on a set of measures to track planning effectiveness and then conduct such a program consistently over time.

Studying deviations from plan is an ideal aid to management direction and tracking progress and is an exceptional tool to help control subordinate company functions, too. List all relevant issues or problems (here, deviations from plan), prioritizing the items. There usually is more than one possible ranking, but one sound criterion is ultimate company impact, often best assessed by reckoning incurred costs. Let's follow the latter wisdom and rank the problems according to relative cost impact on the company. Now, tackle the problems in top-to-bottom priority order, thereby addressing the biggest issues first. Usually, the top 10–30% of the problems carry 70–90% of total impact, a basic concept dubbed the *Pareto principle*, or the 80/20 rule. Remaining issues can be cleaned up later, in an environment of less urgency or inefficiency. For the Pareto principle to really work, management must dedicate specific commitment and resources to the program sufficient to accomplish true progress.

Finally, using comparisons of business plan versions to evaluate planning methods can be quite valuable. Think long term. Planning must be integral to company action through its entire future tenure. Continuously reevaluate planning methodology and explore means to improve it.

6.1.7 Departmental Plans

Departmental plans are detailed specifications for present or future policies, procedures, or actions, primarily focused on one department or company functional group.

Departmental plans are not mandatory in every case, in contrast to the formal business plan for the whole operation. Management may deem the business plan proper as stipulating all necessary detail to completely determine company policy, procedures, and actions. In other cases, the business plan may not fix all company elements to a great enough degree for efficient C^3. In such cases, departmental plans can be invoked for some or all company functions to define behavior with more rigor than the business plan. Alternative approaches to departmental plans might rely on the management team to interpret and enforce business plan dictums, imposing formal companywide SOPs or creating highly specific operational manuals for specific company activities deemed in need of extraordinary control.

Departmental plans are therefore one factor in the arsenal available to impose C^3. Every project or company is different, having a distinct product line, unique management makeup, and a set of physical and behavioral norms all its own, the collective of all norms forming the company culture. In any case, departmental plans can form a significant component of C^3, and all departmental plans must be entirely consistent with the business plan. That may sound obvious, but there have been innumerable instances in which business plans concerning individual functional subdivisions contradicted the company's overall business plan.

6.2 Structure of the Business Plan

Having completed a tour of the primary purposes of a business plan, we now lay general background for typical business plans suitable for the privately funded tier of project complexity. First, we divide the structure of the business plan into three primary sectors:

- *Narrative.* The narrative is the main text of the business plan and often begins with some form of summary or abstract of the global plan proper.
- *Financial model.* The model, as it is called, is typically a multiyear financial spreadsheet that relates estimated revenue and cost elements for each of many time periods.
- *Appendixes.* Important supporting data are usually attached as one or more appendixes, so interested readers can review specific, detailed supporting data relative to more concise descriptions presented in the narrative or model.

The following subsections address, in turn, each of the three major business plan sectors. Appendix A provides significant additional detail and lays out key examples and actual implementations.

6.2.1 The Narrative

The narrative is the main text portion of the formal business plan. It may include graphs, tables, photographs, or other auxiliary material, but by far the main form of expression is textual.

The purpose of the narrative is to paint a verbal picture of the project or company, the more vivid the better. The goal is to treat every topic of recognized importance with enough detail and rationale to fully describe the main thrust and important nuances of the project and its prospects. The typical, nontechnical reader should be able to comprehend easily the project's complete history, status, and future plans.

The following list sketches the main topics traditionally taught for building a professional business plan. The data have been extracted from real-world business development, consisting of impressions from tens of books and various educational materials, plus hundreds of business plans encountered over a decade. The material is also based on discussions with the authors or representatives of an estimated 20–30% of the observed plans. This outline of a business plan is thus a composite of a large number of individual plans or primers, intended to show main topics and order of presentation.

- *Table of contents*, a list of all sections and subsections, by page;
- *Summary*, an overview of the contents of the business plan;
- *Introduction*, a project perspective, including general strategies;
- *Product(s)*, a detailed description of the key product(s);
- *Marketing*, a usually shorter treatment of marketing and/or sales, including competition;
- *Manufacturing*, a typically brief discussion of how the product will be produced;
- *Management*, a list of key people and their titles and responsibilities (resumes included here or placed in an appendix);
- *Financial projections*, the financial spreadsheet: an income statement, sometimes a balance sheet, and rarely a cash flow analysis.

The sources of this amalgamated outline show considerable variability in exact topics and ordering, a pivotal issue that is developed next.

It is difficult to write a formal business plan, and an outline alone does not provide a solution. Plan prowess also depends on preparatory business planning, knowledge of business development, and, not the least, writing ability. Software packages for do-it-yourself business plans exist, but the structured format can render specific project accommodation weak. There is a tremendous amount of information to be covered, at seeming odds with the brevity sought by potential investors. Many approaches have been tried, and most have often received cool receptions. The problem is extreme document sensitivity, since the slightest misstep precipitates an avalanche of criticism, especially when the plan is distributed outside the friendly confines of the company.

We were no different at the author's business development firm, Venture Solutions, Ltd. We responded to early critiques with myriad experiments of substance and style. We did have an advantage, having seen literally hundreds of other plans over time. In any case, in testing many, many ideas, we at VSL finally developed a format that seems noticeably superior:

- *The deal*, a very brief description (two or three shortened lines of data), placed at the head of the executive summary, and centered;
- *Executive summary*, an overview of major points of interest to investors, carefully limited in length;
- *Table of contents*, the same as the traditional table of contents;
- *Program overview*, a novel section with a unique purpose, designed to counter certain plan objections that we at VSL had encountered regularly;
- *Management and strategies*, far beyond a team description, an examination of business development strategies and much more;
- *Products*, product function, design, and protection, and the nature and direction of future product development;
- *Marketing and sales*, detailed treatment of customer analysis, program design, and planned product imaging, positioning, promotion, pricing, and delivery (all marketing), sales strategy and tactics (sales), and customer service (marketing);
- *Competition*, an analysis of the major products, their advantages and disadvantages, and how the product addresses those issues;
- *Operations*, a concise but detailed summary of all production, purchasing, inventory, shipping, and returned goods activities, including quality control;
- *Risks and rewards*, a special section to highlight all important investment risks, how the risks are being addressed, and potential rewards for company involvement.

Our experience revealed that the described outline exhibited successful solutions to major problems, solutions that obviously were not in common practice. After developing the outline, we used it in writing several years' worth of business plans for a myriad of projects and carefully observed how various documents were received by a large, diverse audience. Prior to implementing the above outline, our plans were rated quite good, but then the same people went on to list one or more criticisms. Not everyone mentioned the same problems, so the list was of some length. Thus, our early business plans were compared favorably to most others, yet they contained perceived weaknesses, at least to some people. Subsequently, when we introduced our new outline, the results were astounding. Suddenly, the criticisms almost vanished. Appendix A provides a detailed review of our business plan outline and its implementation.

Notice the profound distinctions in the topic order of our outline, compared to that of the generic business plan list. The following subsections address our business plan outline, explaining the rank order, content, and rationale for each item.

6.2.1.1 The Deal

Above the start of the executive summary text, we place a centered, heading-like element, two or three shortened lines long, stating main terms of "the offer." This applies only to plan versions that, among many purposes, include a search for funding. Other plan writers may have used such a highly visible display of the offer as well, but we have never seen such an occurrence. We installed the offer placard to counter historical comments that the offer is hard to find or absent from most plans.

6.2.1.2 Executive Summary

The executive summary is a concise abstract of the business plan's main narrative, ending with a four-line table that lists sales, investment, market share, and pretax profit amounts for each of the plan's 5 years. The executive summary, including the offer and the ending table, should never exceed $1\frac{1}{2}$ pages in length.

6.2.1.3 Table of Contents

This item is essentially self-explanatory. Our approach is to show not only section (chapter) titles but also first-level subsections under each section title and a numbered list of appendixes and their titles. Beginning page numbers are indicated for all listed items.

6.2.1.4 Program Overview

The program overview is undoubtedly our most pronounced business plan innovation, having induced the most striking comments and results. Besides the 50 or so plans we have written ourselves and the hundreds reviewed, we have seen some semblance of our program overview technique used only a few times, so elaboration is warranted.

Assume that an exemplary business plan has an executive summary of $1\frac{1}{2}$ pages and a main narrative of 20–30 pages, numbers reasonable for the complexity of a privately funded plan. The first of seven sections in the main narrative is the program overview, a 4- to 7-page expanded summary that is still far shorter than the main narrative. Comparing lengths, the narrative is roughly four times the overview, which, in turn, is around four times the executive summary. Put another way familiar to our technical audience, we created an entirely new section roughly halfway between the brief summary and the detailed narrative on a logarithmic (not linear) scale.

Real-world feedback has shown the program overview to create much more versatility and efficient functionality for a wider audience. For anyone reading the whole plan, the staircase presentation allows reinforcement of key facts. For mixed groups (e.g., different departments or tiers of authority), it provides an extra rung of descriptive complexity. Some readers need only a cursory look for orientation or to determine next routing (the executive summary), others only an intermediate treatment (the program overview), while some need total document comprehension. Our design at once meets the spirit of a multi-element description set. Added discretionary features include a frontal explanatory page to outline the unusual document structure and tabbing or color-coding major business plan sections for easier navigation.

Incidentally, our program overview content is novel, too, essential to avoid redundancy with the shorter summary and the longer narrative. All three major text tiers in the business plan (summary, overview, narrative) address the most important issues of the company, but from a different perspective. Examples of distinct treatment follow:

- *Funding*, an explanation that equity funding is needed and a list of major funding uses (summary); tabular description of funding plan, segregated by business development stage (overview); and investment risks and rewards, addressed as the last full text section (narrative);

- *Target market*, a brief description of the overall marketplace (summary); a main development of market size estimate (overview); and, using the previously derived market size estimate, a primary development of market and market share estimates, spanning the 5-year plan (narrative);

- *Inaugural product*, a brief description of the initial product, its competition, and its most prominent advantages (summary); a discussion of product nature and uses (overview); and a comprehensive treatment of the product in a standalone major section (narrative);

- *Competition,* a brief description of competition treated collectively (summary), a brief summary of competition addressed individually (overview), and a comprehensive treatment of the subject in a stand-alone major section (narrative).

Careful redundancy management affords reinforcement of important issues without repeating the same information.

The program overview is placed as the first section in the main narrative for two reasons, optimal accessibility and as a natural transition to introduce the remaining main narrative. We defined an introduction section at the head of the main narrative in the traditional generic business plan outline. Almost every plan we have seen begins with some sort of introduction or transition. In our business plan outline, the program overview replaces the generic outline's introduction section.

6.2.1.5 Management and Strategies

Notice that our business plan outline places the management discussion first (or at least immediately after the program overview), consistent with its general ranking by sophisticated investors and dramatically contrasting with the ending placement (right before the model) in the traditional plan narrative. We also have incorporated the main discussion of overall strategies here. Someone following the generic plan outline probably would treat strategies in the introduction or the product section. Our rationale was to combine management and strategies in a way that highlights the company's administration function, since it is management that develops and implements the strategies.

6.2.1.6 Product

This section is fairly self-evident. It begins with a product introduction, design approach, detailed description of the product and its various functions, and the resources required at various stages to accomplish product development and design. Also addressed are the patent, trademark, and regulatory status of the project. The product section closes with a look at additional products (or at least rough product categories) envisioned for the future and the nature of any ongoing or planned research program for new product ideas long term. These ending items are crucial, because one-product companies are not attractive in the privately funded sphere.

6.2.1.7 Marketing and Sales

Detailed in this section is the general marketing plan (as previously summarized: customer analysis; program design; planned product imaging, positioning, promotion, pricing, and delivery; customer service). The sales plan is also discussed in some depth, outlining how sales are to be accomplished and who is responsible. The section closes with prime data to establish or estimate the size of relevant markets, projected market share versus time, and a description of ongoing marketing research that will be in place. The market size and share are paramount topics that often are omitted from plans. Readers want to know the basis of model revenue figures and how much market share such figures require.

6.2.1.8 Competition

This section is meant to scrutinize the company's main competition. The analysis should be as quantitative as possible and should include both individual competitive products and also some detail on the companies behind the competing products. Market share estimates should be justified from a technical standpoint here, by showing how your product's attributes would be expected to stack up against the competition and the resultant impact on product perception and sales.

6.2.1.9 Operations

Operations are defined as those activities associated directly with development, design, production, delivery, and quality of company products. As with the competition section, the presentation must be as quantitative as possible to facilitate a thorough description of operations, but it is limited typically to no more than two pages. Key issues are how the listed activities are structured administratively, who is in charge of what activity, and the raw basics and rationale of how each listed activity is to be conducted.

6.2.1.10 Risks and Rewards

This is a special section that we have added to all business plans. It imparts major clarity for an issue that is often omitted or slighted. Two topics are included. First, all major risks are identified, and steps taken to counter each listed risk are explained. Second, major benefits for stockholders, officers, and employees are identified, to concisely show how these various benefits are distributed.

6.2.2 Modeling and the Financial Model

For this book, we define the model as a multiyear financial spreadsheet that relates estimated costs and revenues for each of many equal time periods.

For all modeling, we uniformly use 5-year durations, the business development consensus for product-oriented, high-growth, privately funded companies. Regarding time intervals, many formal sources stipulate that every business plan year must be modeled by month. Other sources may say that after year 1 or 2, statements can revert to quarterly intervals, monthly spans being required until then. We have informally tested time interval parameters in 30 to 40 business plans, and the results indicated that quarters are fine for every year, even the first. We subsequently substantiated that conclusion by testing large numbers of real business plans, and we have never heard so much as one negative comment. Most plans probably can universally use quarterly time periods with no problem, although there may be times when gyrating project dynamics (e.g., large seasonal swings) call for tighter intervals. We recommend that management use their best judgment on the time intervals or pattern of intervals to employ.

Chapter 5 outlined the three main financial model statements, the income statement, the balance sheet, and the cash flow statement, which, taken together, characterize a model. All three statements must cover the entire business plan interval. The income statement is a summary of all company inflow (revenue) and outflow (expenses) for every relevant quarter, the balance sheet displays total company assets as equal to the sum of total liabilities and stockholders' equity (the latter being an analog of an individual's net worth), and the cash flow statement specifically summarizes the movement of cash in and out of the company, as a combination of past activity (actual transactions) and future projections (estimated transactions).

We use certain techniques to simplify the look and feel of a model. One way, as already mentioned, is to use quarters for the basic time interval, unless meaningful evaluation dictates finer resolution. Another is to restrict every financial statement to just one standard 8.5-by-11-inch page, an approach that has received many compliments from people who have struggled with, say, an income statement that fills two or more pages. The disruption of incessantly turning pages to explore the statement is a trying experience. We find that combining the use of quarter-year columns, an efficient roster of rows, and the "print to fit" feature of popular spreadsheet software creates a nice solution. Full 5-year financial statements can be printed on one page in a font that is small but readable. We have heard few negative comments, certainly fewer than those about the often seen technique of oversized pages or multiple-page statements.

There are two extreme philosophical ways to compute a business spreadsheet model. One is an attempt to define actual costs for every item and activity, what we call the *actual cost* method. Its opposite is the *percentage cost* method, in which costs that are not known very well are proportioned. For example, we might define R&D, facility, or any other costs as a percentage of total expenses or as a percentage of sales revenue after such money begins to flow predictably. Sliding-scale percentages might even be used to account, say, for lower sales revenue relative to costs just after product introduction compared to later in the product's sales lifetime. By using a percentage approach, you can in essence restrict selected activities to a defined relationship compared with other expenses, revenue, time of year, or many other factors.

Merits of the actual cost versus percentage cost methods of modeling vary with company operations. Where most if not all costs are known or definable, the actual cost method is desired for improved accuracy. That could occur, for instance, when the startup company employs well-known technical and marketing operations, such as a new product class within an established industry. Usually, though, actual costs are not known accurately, because development times and costs for various functions are uncertain, and ultimate sales are unpredictable too. An example would be a startup company whose product opens a whole new market or involves difficult design or production issues requiring substantial development for resolution. The best that the management team can do is render careful educated projections, and it may be just as accurate and a lot easier to use the percentage cost method.

Many times, the income statement, the balance sheet, and the cash flow statement are not enough. Because, for reader convenience, we decided to limit each statement to one page, data must be concise. Detailed breakdowns are not possible, because there is not room. Nevertheless, as shown in Appendix A, we recommend that supporting data be provided whenever significant for reader understanding and interpretation.

This seeming paradox is resolved by adding one or more extra financial statements to the model (Figure 6.2). The three main financial model statements remain the same and are considered to make up the primary financial model. We recommend adding several supporting statements that together constitute the secondary financial model. Again, each supporting statement should be designed to fit on one page, just like the primary statements. Appendix A covers the nature and content of a recommended package of supporting financial statements, the basic idea being to provide detailed breakdowns for a number of primary financial statement areas where data are highly condensed. In that way, the business plan reader can choose between summary and detailed data.

6.2.3 Appendixes

Appendixes allow adding relevant business plan supporting material that does not blend well with the narrative. In that way, important but disjointed material can be bound within the business plan proper. Appendixes are consecutively numbered using a unique system (e.g., roman numerals) and placed after the main narrative.

Many plans include the financial projections as the main narrative's last section. While somewhat academic, that is really improper. The model is technically not in narrative form but is primarily a spreadsheet, with perhaps some introductory or explanatory narrative (such as a list of primary assumptions).

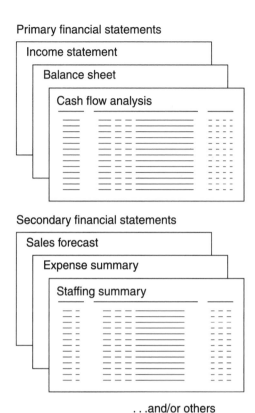

Primary financial statements

Income statement

Balance sheet

Cash flow analysis

Secondary financial statements

Sales forecast

Expense summary

Staffing summary

. . .and/or others

Figure 6.2 Primary versus secondary financial model statements. The complement of primary financial statements is invariant. The complement of secondary financial statements is discretionary, but the specific examples shown provide a solid and logical set (see Appendix A). Each statement is designed to display all 5 model years on one page oriented in landscape configuration.

Unless compelling problems result, it is best to be technically correct. We avoid placing the model in the main narrative but position it as the first two appendixes instead. The first and second appendixes thus present the primary and secondary financial statements, respectively. We separate the primary and secondary statements so readers can more easily locate the desired level of treatment, but the entire model could just as well be placed in one appendix. Given the overriding significance of the financial model, it should always come first in the succession of appendixes.

Appendix A describes a spectrum of appendixes appropriate to diverse business plans. Any one project may not use many of the samples or may even go beyond the provided list. Every project is different, and the appendix list should be designed to cover specific supporting material that is available and important but not appropriate for the main narrative. We have seen excellent, complete plans that featured only 1 or 2 appendixes, while others may have 10 or more appendixes without seeming artificially extended.

6.3 Assembly of the Business Plan

The best embodiment of a business plan is a single bound volume, with the following elements:

- *Covers.* Use a heavy, high-quality cover material, which can be obtained at local specialty stores or large chain office supply firms. Professional visual appearance and feel are critical.

- *Binding.* Many professional binding systems exist, some virtually equivalent to quality books. Still, our tests reveal superior plan utility comes from using a garden-variety plastic comb binder. That allows the business plan to be naturally laid fully open at any page, without the complaint-generating need to pry and hold the business plan open with often significant force. Three-ring notebooks work too, but we found it difficult to attain a satisfactory professional look and feel.

- *Sections.* We use dividers to segregate all major document sections, including the executive summary, table of contents, each narrative section, and each appendix.

- *Order.* We recommend using the exact structure, outline, and order of sections described throughout this chapter and Appendix A.

- *Length.* Targets for the executive summary and the entire narrative (including the program overview) should be 1 to $1\frac{1}{2}$ and 20 to 30 pages, respectively. Midway between those two on a multiplicative (logarithmic) scale is the program overview (4 to 7 pages). Appendixes can vary greatly in number and length, but each should be relevant and concise.

- *Quality.* The business plan should exude business quality but not be pretentious. Use high-quality paper, tabbed section separators, and a high-quality, attractive cover that has the look and feel of an important business document.

- *Serialization.* Apply a unique serial number to every copy of the plan and have each recipient sign a nondisclosure agreement (see Section 6.5.3) prior to receiving a copy. Keep a log that tracks recipients, serial numbers, dates of issue, and expected dates of return. Regain possession of every plan (investors or professionals can be allowed to keep one copy).

These are not hard and fast dictums, but experience argues for close adherence. Often the business plan represents first company contact with a prospective professional or investor; if presented wrong, it easily could be your last contact as well. Manage business plans correctly, thereby controlling one area of potential concern.

6.4 Alternative or Contingency Plans

Planning basically is a forecast of the future. However, hardly any plan is perfectly realized in practice. Many types of purposeful actions by competitors or other entities and unintentional effects from the world at large can wreak havoc, the best laid plans going far awry.

If the company business plan is critically dependent on one or more key assumptions, contingency plans may be wise. One way to handle especially critical assumptions is to produce more than one plan, allowing a detailed plan to be in place regardless of which way pivotal eventualities go.

Contingency plans are not for everyone. The most important condition to evoke a contingency plan approach is the existence of, at most, two or three prime assumptions that can go only two or three ways. Even with three assumptions and three paths each, nine plans would be required, already beyond the normally practical. In fact, given a situation that unavoidably involves multiple paths, first try to reduce the possible paths through creative planning. However, when all planning options seem exhausted, contingency plans can be both useful to the company and impressive to outsiders.

The most direct way to develop contingency plans is to create multiple variations of one basic financial model. One unique set of column and row headings should be used, so that different model versions are purely distinguished by core data. Then, for items involving critical assumption(s), create two or more model versions that represent each possible assumption path. For each version, you also must adjust other model elements in accordance with the subject assumption(s). For example, if version 2 involves twice the cost of a key part, then version 2 must reflect altered sales projections, inventory costs, quality inspections, and so forth. You probably would include all model versions in the business plan, each separately labeled to make clear the variation of assumptions employed. The exemplary plan would have just one narrative, somewhere clearly describing the tricky assumption(s) and how management would select the appropriate model version once problematic outcomes become better known.

6.5 Product and Company Protection

During the business development process, progressive product and company advancement builds value, which needs to be protected. Five types of protection are of import: patents, trademarks, copyrights, operational agreements, and nondisclosure agreements.

6.5.1 Patents

A patent provides its owner the exclusive right to make, use, and sell an entity or process for 17 years. Patents are acquired by submitting an application to the Patent and Trademark Office of the U.S. Department of Commerce in Washington, D.C. It is possible, but not advisable, to prepare and submit your own application. Retain a firm that researches and prepares patents and other forms of intellectual property protection.

Patents have both good and bad aspects. Once a patent has been awarded, it is good that the holder is the entity of record owning the patent rights, but the bad news is the government does not defend those rights. The patent holder must bring action against any person or business entity felt to be infringing on the holder's rights, typically a very expensive process. Also, other entities may challenge the patent's basis, attempting to officially reverse its issuance.

The common form of patent that most people envision is termed a utility patent, a class designed to protect an inventor-devised means to accomplish some particular functional utility. A second, much less common class is the design patent, a vehicle granted on a specific dimensional form and ornamental appearance, any functionality provided by the design not being protected.

6.5.2 Trademarks and Service Marks

This is a distinct protection granted by the Patent and Trademark Office and by many states. A trademark provides the holder the exclusive right to use a rather precisely defined symbol, name, designation wording, or other representation. A trademark is quite specific, characterizing not only the wording, if any, but also the graphical portrayal of the mark, including its color, detailed content, and any other distinguishing features. Trademarks are commonly used to protect public marketing insignia or designs.

Service marks are basically a nonproduct analog of trademarks issued by the Patent and Trademark Office. Product-oriented companies use trademarks to protect such things as company name, logo, and specific product or product family names. In contrast, firms whose main activity is services protect the service names or designations and associated graphics and logos with appropriate service marks. Companies with mixed offerings may find both trademarks and service marks useful.

6.5.3 Copyrights

Copyrights are issued by the Library of Congress, not the Patent Office. The copyright laws convey strong protection for the ideas of authors, artists, publishers, and people who produce certain other forms of craft. A huge, relatively new avenue of copyright protection involves computer programs, the legal view being that orchestrating digital pulses or other signals to carry out useful functions is subject to copyright coverage.

Software developers find both patents and copyrights applicable. Patents apply in realms such as program algorithms, display presentations or arrangements, menu layouts, editing functions, compiling techniques, program language translations, and operating system methods. Also, design patents are occasionally used to protect a particular computer screen portrayal. Copyrights are employed to protect software against duplication, creation of derivative versions, and distribution to the public by sale, rental, or other means.

6.5.4 Operational Agreements

Operational agreements are formal company arrangements with an outside entity controlling one or more forms of intergroup action. These agreements should always be in writing, no matter how routine or simple, and should meticulously spell out what is expected of each group. The latter point must include recipient protection of all exchanged information and tangible elements, such items being proprietary to the originator.

6.5.5 Nondisclosure Agreements

Nondisclosure agreement is the short-form designation for a nonuse, nondisclosure, and noncompetition agreement, used to restrain a party's ability to use or disclose information or knowledge provided in agreed confidence or to compete against the granting entity. Two general types of agreement are relevant to the privately funded, high-growth, product-oriented company, depending on the nature of the second party:

- *Outside entities.* In this case, the purpose is to control disclosure of information provided by company representatives to outside vendors, professionals, consultants, potential investors, or any other person or other legal entity.

Statement of Confidentiality and Non-Use

_____ , hereinafter referred to as "Discloser," possesses certain concepts, ideas, and technology related to the following:_____

The undersigned recipient of such confidential information, hereinafter referred to as "Recipient," agrees to accept such confidential information upon the following terms and conditions:

Recipient agrees to hold in confidence and not to use (for purposes other than evaluation) or disclose directly or indirectly to any persons outside of Recipient any such confidential information (whether patentable or not) for three years, without the prior written permission of Discloser, and further that all the employees, agents, officers, directors, consultants, and associates of the Recipient to which the subject confidential information is disclosed for purposes of the present review shall be fully bound to all conditions of the present statement of confidentiality and non-use.

It is understood that the following information shall not be considered as confidential information:

A. Information that, at the time of disclosure to Recipient, is in the public domain by written publication;

B. Information that, after disclosure to the Recipient, becomes part of the public domain by written publication through no fault of the Recipient;

C. Information that Recipient can show by written documentation was within its possession at the time of disclosure hereunder to it and was not acquired directly or indirectly from Discloser; and

D. Information that Recipient can show was acquired after disclosure hereunder to it from a third party who did not receive it, directly or indirectly, from Discloser and who did not require Recipient to hold such information in confidence.

Recipient agrees to accord to the confidential information the same degree of care and use the same confidentiality practices as Recipient exercises or employs with respect to its confidential or proprietary information.

Upon request by Discloser, recipient will return to Discloser all document originals, document copies, and other tangible property that contains confidential information.

The discussions referred to in this agreement contemplate that, should the parties decide to proceed with the use of the idea(s) or information, a detailed agreement will be prepared and signed, delineating each party's respective rights and responsibilities.

By signature below, Recipient warrants full authority to carry out the terms and conditions of this agreement.

Date   Recipient's Name (Person or Firm)

By: (Representative of Recipient, If a Firm)

Title of Recipient Representative

Figure 6.3 Complex nondisclosure agreement. The simple version is the same but without the fourth paragraph ("It is understood that"), including the four-item list. The simpler the text, the more protection is afforded the discloser. Exclusions weaken the discloser's position relative to the recipient.

- *Employees.* Companies often require all employees to sign an agreement controlling use of company information during and after company affiliation. Often, clauses are included relating to ownership of inventions an employee creates while employed by the company.

Agreement applicability, scope, details, and implementation requirements can have local nuances. Consult appropriate legal counsel for proper need, design, and use.

Relevant to business plans, every person or group from outside the company provided a business plan for review should be required to sign a nondisclosure agreement of the outside entity type. Figure 6.3 is a complex agreement for use with outside entities. Very simple agreements are better if you are the party seeking protection, because every added clause typically excludes something (provides a circumvention opportunity). If you are the one signing, it is best to add clauses to clarify issues of concern.

6.6 The Business Plan As a Living Document

We have emphasized that the business plan is not a one-time endeavor. The primal flash evolves into a product concept that eventually spawns a business concept as both continue to develop. Because the business plan is merely a company snapshot at one moment, it must undergo multiple revisions as time passes. The business plan is thus said to be a living document that must be attended.

The business plan is an enigma to many; they cannot understand its absolute need or the tremendous advantages it can offer. Properly write the business plan and keep it current, and you will be amply rewarded over and over.

This chapter has emphasized that the business plan's utility in the funding process is only one of its many benefits. Still, funding is without question an issue of monumental importance. Notice that the rationale for many aspects of the business plan is closely tied to funding. The process of acquiring funding is incredibly demanding, and the business plan is designed to smooth as many rough spots as possible.

Close juxtaposition of business plan and funding discussions is useful, because the two subjects share significant although often subtle considerations. In that spirit, we can now focus on an in-depth look at the frequently fearsome hurdle of funding. Done right, it may not be so bad after all.

7

Funding

7.1 Short- and Long-Term Funding Plan

Funding probably is the least understood aspect of business development. Even so, many people enter the process with a smug arrogance, based on a self-expressed feeling that the project is so exciting it will sell itself. Others see the funding process as a totally mysterious realm, populated by a select club of the wealthy, knowledgeable, or well connected that hardly ever accepts newcomers.

Both groups are wrong, but the latter crowd is closer to the mark. The funding arena is distinct, at least in degree, probably because its loose, informal domain has evolved a unique language and rules and has enjoyed an extended existence with ample opportunity to accumulate convention and tradition. Funding deals with a consistent set of facts and issues, standards of presentation, and other rules and patterns of engagement.

At any point in time, one can divide the funding problem into two components, short term and long term. For illustration, the short-term horizon includes business plan maintenance, the current round of funding, and candidate investor presentations, while future funding considerations, stock allocation plans, and 5-year strategic planning represent long-term elements. There is no absolute rule dictating the length of the short-term factor, but short-term is frequently taken to mean annual, as in strategic planning. For our purposes, the short term also loosely correlates with the current business development stage under discussion.

People's attitudes and actions are often quite distinct relative to short- versus long-term funding. Short-term funding is seen as an immediate need required for current work to proceed, while long-term funding is a future issue of little current impact. Such an attitude toward long-term funding is seriously flawed, because it is essential to carefully allow for long-term capital in the process of finding short-term funding. The best view is to consider the work supported by each funding round as primarily directed to proper company posturing for subsequent funding. Short-term funding is an active, ongoing process, carefully positioning the company through the passive allocation process of long-term funding management for optimal investment attractiveness.

7.1.1   Short-Term Funding

We have declared short-term funding to be an ongoing process. The funding is needed now, within one to four quarters, installments possibly being acceptable. Even so, later portions of short-term funding require early commitment, because development of a product or an entire company depends on a continuous, interlaced series of steps.

As an illustration, if steps 1 and 2 are funded, but later short-term steps are not, failure of later installments to materialize can cause disastrous work stoppages. Launching work before obtaining commitments for all short-term funding is risky, but at times there is little choice due to some window of opportunity or necessity (e.g., seasonal or economic cycle-dependent raw material or school-year student labor). Usually, though, it is best to delay the start until all short-term funding is committed, to protect both project and investors. The project is better off due to a clear mandate and resource base for planned work, and investors are better off for the same reason, having better assurance the project will reach the next milestone. Each funding round and work cycle completed positions the project with more strength and value, due to further advancement. In turn, that leverages future-stage funding, due to more tangible project evidence such as planning, designs, and prototypes to show potential investors.

7.1.2   Long-Term Funding

Our view of long-term funding is of a passive allocation process. It is passive rather than active because action eventually will be needed but is beyond the short-term horizon. Long-term funding is also an act of allocation, because an appropriate company share must carefully be preserved to cover future investment predicted by the company financial model.

The following schedule depicts a stock ownership allocation plan for the company through all business development stages required to achieve profitability. A specific illustration is presented in Figure 7.1.

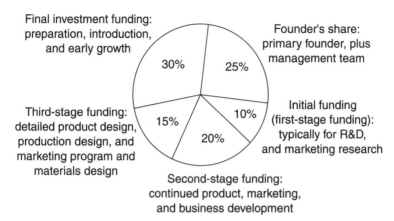

Figure 7.1 Sample stock ownership allocation plan. Percentages and the number of stages vary dramatically across projects.

- *Founder's share* is the company share allocated for the founders (founder plus full management team) (20–40%).

- *Initial (first-phase) funding* is the first funding independent of the founder and the management team, usually applicable to R&D-stage activities (10–20%).

- *Second- through (N − 1)th-phase funding* is the funding that leverages initial R&D and that is used to complete development or at least to advance it to a major new plateau of demonstrable results and resultant funding leverage (10–20%, each phase).

- *Nth-phase funding* is the funding for the phase of advancement that achieves design completion, production readiness, and initial marketing preparatory work (10–20%).

- *Final investment funding* is the funding allocated for product introduction and early growth, including major marketing and sales program design and installation, technical wrapup, production working capital, and early production and sales, positioning the company as able to obtain receivables financing initially and eventually commercial operating loans (15–40%).

- *Commercial loans* are receivable financing and commercial operating loans from financial institutions (0%).

Time frames and company share ranges have been left vague and funding amounts have been omitted due to extreme interproject variability. The relative distribution of amounts across distinct funding phases can vary enormously across projects as well.

Overselling company shares short term is a serious possibility, not an academic musing. We have seen projects sell too large a share early and become trapped. The short-sighted reasoning is that the original pie can be made larger, if need be, to allow added investors later. The rub is that, given such action, all original investors become diluted, a situation you and even your team may find acceptable to maintain advancement, but the original cash investors may not. They could reasonably argue that their stock purchase reflected a promised company share, one they expect to keep, triggering company-killing lawsuits that never should have happened.

Besides share oversell, share undersell also has been seen. The company share is too small for the respective investment, compared to typical projects. When investors pay too much for early company shares, the project can later face a no-win situation, as shown by these two outcome extremes:

- *Second-round funding at normal share prices.* If you are able to secure second-round funding but at a normal share price (rather than at another undersell), original investors will be furious and for good reason. The original investors got far less company share per dollar than later investors, exactly counter to normal reasonable expectations of lower share price to offset greater risk of earlier participation.

- *Second-round funding at another undersell.* Such a project is precarious. One or more consecutive undersells imply a greedy founder trying to retain excessive ownership. In general, investors vary in sophistication, but most understand and expect general fairness. Eventually, the underselling project will hit a point where no funding can be found, the only cure being complete project restructuring, with the founder taking a much smaller share and everyone else being repositioned fairly. Underselling founders seldom face reality soon enough to avoid collapse, and all prior investors go down with the captain and the ship.

The answer to all that trouble is simple. Do not oversell or undersell a company. Right from the start, make sure the pie (the company) is fairly allocated to allow for all required funding, long term as well as short term, until profitability is achieved. This is a sterling reason for dutiful and very early professional planning, since you must account as best you can for all future funding needs even before the first dollar is received.

The attempt has been to place short- and long-term funding into realistic context. Notice that presently required funding is always short term, regardless of the round of funding (first, second, and so on), whereas initial funding occurs but once and the final funding round has no long-term component. This is the last we will say about long-term funding, the message being to always fully allow for this most critical strategic consideration. Our focus now shifts to detailed considerations of immediate investment pursuit, which is, by definition, always short term in nature.

7.2 Strategy and Tactics for Initial Funding

Initial outside funding happens once, being the first funding sought external to the management team. It is lucky when building the management team leads to significant funding as well, even luckier when the founder alone can carry the project far down the road. Remember, though, luck is not an authentic business development tool—to depend on luck is usually to lose control of the endeavor.

The winning combination is hard and smart work, toil alone being insufficient. Working smart consists of thoroughly knowing your product, your business, and the market and then orchestrating proper funding. On the basis of your product concept, you build a business concept to enter a targeted market. That plan, plus a potent management team, a top-level business plan, and keen savvy of the funding process, is your ticket to funding.

Funding cannot be guaranteed. It is possible to completely fulfill all the above listed attributes and still not find funding. However, if all company attributes are at hand and truly strong, and the project makes excellent economic and marketing sense, funding almost certainly can be found. This bold assertion assumes persistence, the attitude and endurance to never take no for an answer. When you have a winner, you have made it look like a winner, and you keep maneuvering the endeavor into the right position for a win, odds are you eventually will triumph.

In that spirit, initial funding is no different from any other funding stage. With the right tools and techniques in place, the funding should happen. Still, initial funding has unique aspects, even if mostly in degree rather than kind. This chapter addresses several key funding elements and their particular initial funding nuances.

7.2.1 Having a Completed Business Plan

Do not start any aspect of the funding search until the business plan has been completed, not just written, but printed, bound, and fully ready for distribution.

It is tempting to begin discussing the project early, since excitement and anticipation are high, and it is natural to talk about what is on your mind. Even so, danger lurks. There have been countless projects where the founder or an associate approaches someone and tells their tale, the prospect becomes interested and wants to learn more. What may seem good is really a potential disaster.

Demand has been created, a call for more, but no more is available because there is no business plan, and the rest of the project is just ideas. At such a point, the business plan is the primary selling tool, and nothing, not even verbal discourse, compares. In this situation, virtually every prospect has been lost across all projects that we have observed, the prospects seeing grossly unprofessional behavior for what it is.

Potential investors have many elements in common with ordinary shoppers. When learning of the project, they want all pertinent data right away. Emotions are fickle and impermanent; once you get a prospect interested, you need to deliver quickly. Remember, you get only one chance to make a first impression, and trying to rekindle initial prospect emotion at a later time is virtually impossible. Have the business plan completely finished before you begin shopping the project in any way.

7.2.2   Having the Project Ready for Showing

Before you begin shopping the project, everything else in addition to the business plan must be fully ready as well.

Projects experience the requirement for an initial funding search at different stages of development. Never begin shopping the project too early. Needs may change as development proceeds, leaving your pitch with improper rationale, amounts, or timing of funds. The preparation must include a clear understanding of exactly how much funding is needed, its timing, the intended use of the funding, and what is being offered for the funds. Finally, you must have the project groomed so that the status of various completed and ongoing activities is professional and can be intelligibly shown.

7.2.3   Organization

Whatever the stage of funding, you must carefully organize and track the effort. Organization is required to look and function professionally in this most critical of activities, every action being planned and tuned to run smoothly and efficiently. Image is everything. Of course the project and the business plan must be in order. Far beyond that, though, you and your team must appear knowledgeable, efficient, and dedicated to the company and the plan. In addition, it is crucial to have all demonstrable evidence of project progression and state of advancement in neat and logical order.

7.2.4   The Search for Funding

The latter call for deliberate organization certainly should be extended to the formal search process itself. The term *formal* is an apt one, because the search process should be mapped out and then conducted like well-rehearsed acts of a play. Search process details are more intricate and extended than a quick mention here could do justice, a topic reserved for Section 7.3.

7.2.5 Follow the Law

It goes without saying that the company and all its actions must be lawful, but here we are tending a different problem.

The acquisition of capital for startup companies is highly regulated in the United States by the Securities and Exchange Commission, which conveys major impact on your endeavor, as detailed in Section 7.4.

7.3 The Search for Initial Funding

All funding stages face the same hurdles and are hard, although the first stage is the most arduous. This section outlines a stepwise process to seek funding, applicable at any stage. Also conveyed is each step's rationale, since knowing both the how and the why instills flexibility to better handle peculiar circumstances.

7.3.1 Lay Out the Plan

Be deliberate about organizing the funding search. Devise a detailed plan that covers every step, taking care to anticipate every kind of reaction that might occur in different situations. Chances are you cannot think of them all, so later surprises are likely, but detailed analysis and anticipation will add great depth and breadth of perspective to your preparation.

The plan should be developed as a detailed written document, a process that helps stimulate new thoughts and also find weak realms that, before, were considered under control.

7.3.2 Verify That All Is Ready

We emphasized this step in some detail in Section 7.2, but a stern reminder is in order due to its paramount import. You must ensure that the project, its people, all artifacts, and everything else of any impact are fully ready for the outside world, a realm the company probably has not encountered before. Your goal is to find a secure relationship right away, but potential investors do not necessarily feel such urgency.

Funding is a buyer's market, characterized by a vast oversupply of business plans chasing a select population of candidate investors. Inexperienced entrepreneurs who think their projects will sell themselves are in for a rude, bruising shock. It is not enough for you to have one or two traits that are desirable or flashy—the competition is too tough. Most business plans would not get funded even if they had no competition at all. You must work long and hard to bring all features to superior status.

In Chapter 4, we talked a lot about how to do self-analysis. This is one place to put those lessons to work. Establish a checklist of the critical items needed to make your project as bulletproof as deemed practical, listing items derived in part from any and all chapters in this book. The following sample list identifies major elements applicable early in business development, but each company should develop subitems under each identified element specific to its particular situation:

- The seven success factors (Chapter 2);
- Building the management team;
- The business plan (Chapter 6);
- Strategic planning (Chapter 5);
- SOPs (Chapter 5);
- Concept development (Chapter 4);
- Initial product development (Chapter 8).

Critical discussion of these issues is spread throughout the book, in addition to the chapters referenced in parentheses. The list should be extended as the project advances, with new elements added as you and other players hash and rehash every product and company nuance.

The list will be long, but far from being intimidating it provides a stunning opportunity. There is no worse feeling than confronting a strong potential investor who exposes some major project flaw (e.g., poor business plan, inferior management team, bad market choice or marketing plan development). The subject list is your best defense, the object being to have analyzed the project almost ad nauseam until real perspective has come to roost. Hardly any project is perfect, and efforts of all calibers may have one or more flaws. Even so, really well-prepared projects have at least identified the flaws beforehand, and, through extended planning, measures have been devised to best treat or counter those flaws or weaknesses, illustrated by the following examples of problems and interim solutions:

- *Unfilled management slot.* Use an appropriate consultant in the interim.
- *Design problem.* Define a specific multistep plan to develop a solution.
- *Weak or absent patent opportunities.* Make use of rapid market entry, strong marketing, and active market share defense.

In turn, the business plan should call out known problems and generally explain management's plans to best mitigate them.

7.3.3 Test the Business Plan

It is sometimes possible to test the business plan document itself and the underlying plan by submitting it to some type of forum for critique. It is hard to be specific, because the types of available groups differ wildly by state and even by city. However, the following three paragraphs mention some generic groups we have discovered. Many other forms of help may exist in areas of the country with which we are unfamiliar.

An excellent start would be to seek a small business development center or the equivalent, often affiliated with state government, state or private universities, or even private nonprofit organizations. Valuable critiques can be obtained, but beware of largely academic operations that have little real-world experience. Remember, you are not seeking a primer on forming a general small business, such as a shop or a store; you are specifically looking for help in finding and dealing with real investors in the realm of privately funded projects.

Another source of help might be entrepreneurial organizations, typically nonprofit groups formed with the good intentions of providing a forum for entrepreneurs and investors to meet. These groups work well if properly managed, but they also attract an army of attorneys and accountants trolling for business. The danger is that the organization degenerates into a club for the professionals to socialize. We suggested to one such group that they lay out tables assigned various themes, where people would sit during the main meeting or congregate during open social hours before or after the main affair. That would provide an easy means for the right people to find each other, useful table topics including perhaps business plans, product development, marketing, sales, funding, legal issues, financial issues, and so forth.

Finally, one last entity is the invention, patent, or product development organization (check the telephone yellow pages). The ones we have seen are a lose collection of real and pseudo inventors that meet ostensibly to help each other, somewhat like a support group. Still, the particular organizations we have seen are pretty thin on help, especially the kind you will need. Be especially alert for individuals who prey on people with good ideas.

7.3.4 Build a Contact List

Having completed the preliminaries, it is the moment (really, more like an interval) of truth. You now must make contacts in pursuit of investors.

The first step is to build a contact list, one that certainly includes people or groups that immediately come to mind but that also targets select visible and talented professionals well beyond previous contacts or acquaintances. It might prove useful to hire a consultant to help with both list construction and key introductions. Actually, there is almost always significant advanced knowledge of a formal funding search, so build the list ahead of time to provide ample thought, adjustment, and various forms of research.

A key technique is networking. Begin attending or joining in activities everywhere you think that exposure to business and professional people may occur in good numbers. Social and business clubs, professional business development people, business attorneys, accounting firms, discussion groups, fraternal organizations, theater groups, critique groups, and investment clubs are good bets, but do not stop there. Talk to people everywhere you go, look in the yellow pages and newspapers for other contact avenues, and just stay alert. We are not necessarily talking about a direct investor hit, although that would be a bonus. Spread the word that the project is active and that opportunities for participation exist. You will find a host of new means of exposure as you implement initial thoughts, and those you encounter will suggest many more. The real benefit, though, is in the inevitable referrals, introductions that ultimately provide an audience for productive funding search interactions.

Begin networking as early as possible, but do not tell people your idea initially. Inaugural networking is intended to develop contacts for future use, so when the time is right to build the funding search list, you are much better armed with prime candidates. Those candidates are of two forms. One group is viewed as intermediaries, individuals (friends, associates, employees, and so on of potential investors) who could provide you with a favorable introduction. The other group is potential investors themselves, encountered directly during networking.

7.3.5   Set Up Tracking

We know first hand that the funding search can be hectic. Actually, the busier you are, the better the news, the good exposure hopefully resulting in one or more takers. However, regardless of the inquiry rate, it is easy to lose track of just where you stand with each candidate.

Management of your contacts is far easier if you maintain proper records. There is no best way to do this, but one essential activity is to serialize the business plan copies. Put a unique serial number on each plan as it is produced and assembled and then maintain a log listing the recipient, date, the timing and nature of expected followup, and any other pertinent information. The log can be expanded to become a contact record, to record all candidate interchange and anecdotes, important because it is so easy to forget where you stand with one or more individuals.

It is critical to track every business plan to ensure its proper long-term disposition. Once productive involvement with any person or group has ended, retrieve the business plan and any other highly proprietary material. You may trust someone absolutely or feel they are not a risk for other reasons. Nevertheless, two crucial points arise. First, someone may do something, either intentionally or accidentally, with the plan or its contents that is damaging to the project or the company. Second, it is just sound business practice to maintain tight control on inside information, the business plan being only one of myriad elements.

7.3.6 Initiate Contacts

Finally, it is time to begin the contact process. The centerpiece is the contact list, which should be prioritized. A presentation sketch or script should be carefully developed as well; in fact, several distinct scripts may be in order. For instance, develop different forms of presentation for phone, mail, and person-to-person contacts. All else being equal, in-person presentations usually are best. However, you may have special knowledge about some contacts that calls for another approach, or there may really be no choice, given a remote location of or difficult direct access to the prospect. Approach a few (no more than five) of those listed in order of priority. A one-at-a-time approach is infeasible due to contact delays, unpredictable review intervals, and review extensions arising from innumerable questions or interfering events. As you proceed with the first group and beyond, periodically analyze completed and ongoing contacts as a basis for improving presentation methods and materials.

In fact, it is prudent to employ a practice list as well. Identify a few contacts to call on first who are not considered real candidates but whose opinions are respected. Not only will the act of trial contacts, meeting arrangements, and presentations build your confidence in the funding process, but several issues likely will arise that have eluded all planning.

The prioritized contact list is not all that is at your disposal. Three additional channels of potential funding are vendors, competitors, and advertising.

Depending on the stage of project or company development, you may have conducted business with many suppliers of components or services. One or more of those firms might be interested in diversifying and may have particular knowledge of the field or niche of your idea. For one, you probably sought them out for some reason related to their specialty, so explore possible ways they might supply resources or funding for a part interest in your business. The same logic of common interests and expertise applies to your competitors. Of course you want to fully protect your intellectual and other property before talking to anyone, as discussed in Chapter 6.

The last investor search channel of merit is certainly not the least, that of advertising. Most attorneys will intimidate you with warnings of all sorts of bad things happening if you advertise the project. After all, the "'blue sky" (securities) laws at the state and federal levels were written with the express intent of rigidly controlling the rules of securing investment for startup projects. Even so, we have found a way to publicize your project anyway. Of course, for your own protection, you should fully assess our method relative to applicability in your local area.

Our method is as follows. Place an ad in what you judge to be the largest newspaper in your area. The classified advertising section of choice is business opportunities or headings placed nearby that also relate to investments or money sources. The ad must be worded specifically to attract "investor-participants," a working partner expected to assume key ongoing company duties in addition to investment. For instance, an ad for design stage funding might read:

> Exciting consumer electronics startup, R&D complete, needs investor-participant (min. $75K) with relevant exec-level marketing/sales experience & credentials for meaningful equity share.

Using such an ad technically does not violate typical blue sky laws. Of course, you might get some calls from pure investors as well, who have no desire for a working role, and you are free to develop such inquiries (assuming the ad is legal) since there is no threat of legal repercussions until you actually offer or sell stock. At that point, by working with an experienced business or corporate attorney, you should be able to devise a legal way to bring any inquirer, even a pure investor, into the group. Of course, the respective stock purchase agreement documenting the transaction must involve means fully protecting you from legal exposure relative to the regulatory agencies and the investor. We have successfully used this exact approach ourselves.

7.3.7   Take Corrective Action

As the funding search process advances, surprises will happen, something in (or not in) the business plan or funding presentation scheme causing heretofore unsuspected problems or issues. Actually, that is good news, not bad, even though possibly embarrassing when the problem is discovered by an outsider. The positive result is finding yet one more problem, thereby creating the opportunity to fix it. You will certainly resolve it—that is absolute. You must either solve the problem or develop a plan to contend with any trouble that cannot be totally removed. Further, if you handle the issue professionally, you may even save the relation with the rift discoverer, having demonstrated your team's problem-solving prowess.

The overall search process is an extremely dynamic planning environment, to be viewed as a great chance to hone the plan. To do the opposite, to regard the period as one of turmoil and obstacles, is a mistake. Business at all stages has its problems. Learn early not only to deal with mishaps but to adopt an emotional set that takes problems in stride.

7.3.8 Follow Up on Contacts

Every funding search contact is valuable. Almost all initial contacts will fail to realize short-term money, but the interactions are meritable just the same.

Benefits from funding search contacts build product and company recognition. A given contact may not invest directly but may talk to others, and one of them might invest. Such secondary contacts may also spread the word, and anyone who learns of the project may help with eventual marketing, sales, technical or administrative needs, or publicity, or invest or find other investors for future funding efforts.

The contact list is invaluable long after the initial funding round ends, so it is good policy to keep initial (and future) contacts informed of progress through phone calls, letters, and other means. Continue to stay in touch with these people even after the given funding effort, since they or their contacts may turn out to play vital future roles. It is remarkable how such good will spreads. Many intangible benefits and advantages can come into play from building a positive product and company image through continued networking.

7.3.9 Persist

You must really believe in your project. If you are hesitant for some reason, resolve the uncertainty before you start the funding search; it will haunt the funding effort until you do.

We have been describing the funding process as tough but manageable. Imagine how hard the effort would be on the second try, after a failed initial run. Starting over is incredibly difficult, since many potential investors are especially leery due to earlier problems. It is arduous enough to find funding on the first try, so do it right the first time.

If you really do believe in your project, that fact will shine through during interactions with the outside world. If the project also stands tall against the seven success factors (see Chapter 2), you are in the best of positions. The remaining challenges really come down to execution.

A key personal trait for successful funding is persistence. The road is long, and the going can get tough. It is sometimes tempting to just let the project go. However, that is not what you are about. If you have brought the project to full readiness for funding, then many awesome milestones are already history and you are within sight of a colossal beachhead. The secret at any instant of ultimate stress is the deep conviction that the project is real and can be brought to fruition.

Persistence is especially critical in the initial funding effort, the toughest stage of funding. There is less to show than will be available at later stages, so it is harder to get investors involved. That is why exquisite preparation is essential. If everything is properly together, you should succeed eventually, a truth you must firmly believe. That deep inner strength, the conviction that the project has merit and has been properly prepared, must be mastered, since that is what will carry you through rough times.

Remember, you are in a select group. Few budding entrepreneurs properly prepare for the funding effort, because they rarely know every business development element requisite to high-percentage funding and usually do not fully appreciate the ones they do know. Alternatively, you have already accomplished the monumental preparation task and done so correctly. In a sense, the funding search itself is downhill.

7.4 How to Manage Potential Investors

Finding potential investors is a definite victory. When knowledgeable people begin to show interest, that is a strong sign even though the job is far from over. Now you must deal head-on with these people or groups in an attempt to win their involvement. This section explores important issues and methods concerning these critical interactions.

Say you have had an initial interaction with a particular funding candidate. The initial exchange, by telephone or in person, eventually resulted in the candidate signing an agreement to accept and review a confidential, registered (serialized) business plan, after which the candidate would call back. The following subsections focus on several sequential or alternative steps that could ensue, particularly followup contacts, responses from candidates, dealing with objections, calling for a close, and the closing itself.

7.4.1 Followup Contacts

As initial interaction ends, it is best to agree on a time frame for business plan review. It frequently happens that an initial interaction closes with the candidate seeming excited and impressed, but then days go by without your hearing a word. The question is how to respond. On the one hand, you do not want to appear pushy or, worse, desperate. On the other hand, you do want to find out the candidate's feelings. Further, if other interest also exists, you need to know each candidate's thoughts, to permit appropriate internal analysis and also to present fair and honest perspective to other candidates.

The key to scheduling a followup call depends on the way the issue was left before. If a specific interval of agreed review time has passed, it is certainly fair to call. The approach, however, should be soft, asking if the business plan was clear and to the point. That usually causes the candidate to reveal just how much review actually occurred, knowledge essential for best adapting discussion to draw out true candidate thought. Often little review has occurred, due to interfering activities or perhaps even disinterest. If the reasons relate to interfering activities or are otherwise not negative, then diplomatically redefine a time frame for review and interplay. If disinterest seems the culprit, try to get a sense of why, drawing out the main concerns or objections. Given a feel for the problems, maybe the process can be revitalized (discussed in Section 7.4.3).

Remember, just as funding candidates are reviewing you, you must scrutinize them, a crucial point that is easily forgotten but extremely important. To be acceptable, investors must judiciously fit your business plan and culture. Do not ever jump at the first money that comes along, a common but disastrous mistake. Dutifully examine each candidate carefully, imagining his or her expected attitude and response if, for example, funding estimates prove inadequate or the candidate disagrees with particular management actions. Compare the surmised reaction to the kind you feel would responsibly serve the company and other investors. Carefully assess working relationships to the fullest possible extent now, including frank discussions with the candidate, since much about the future may ride on your choices.

During the funding search, maintain open contact with all candidates, managing the process to keep each candidate advancing through the process. Being overly aggressive or forceful is not productive. The best approach is to set schedules of review at all stages of interaction, since that creates a sound basis for followup. Keep the link cordial but also diplomatically exert professional insistence to move the process along. If activity with a given candidate stagnates, suggest that if the business plan is no longer of use, perhaps it is time to return it. Never let the plan just sit, because that makes your group look unprofessional, and it certainly cannot contribute meaningfully to a successful arrangement down the road. On the contrary, asking for the plan back subtly lets the candidate know that you are proceeding in a businesslike manner, that company time and property are valuable, and that his or her shot at the project is about up.

7.4.2 Responses From Candidates

Responsive candidates who call back in a reasonable time must be orchestrated through a series of stages designed to maintain open communications channels, retain interest, and build a relationship. We recommend a judicious mix of several actions, organized as a progressive series of encounters tailored to match your sense of candidate strengths and interests, as indicated by the following sample agenda for potential investors:

1. *Respond to advertisement.* In response to a business opportunity ad placed by the company, a potential investor calls and is directed to the designated person for funding inquiries. That person describes the project, makes sure the prospect is qualified, and arranges a meeting of the two parties (at the company facility, if there is one).

2. *Meet the designated presenter.* The prospect meets the presenter at the company facility, and they assess each other and review the project verbally. If interest is real, the prospect is required to sign a nondisclosure agreement before further proprietary material discussion. Finally, the prospect leaves with a serialized copy of the business plan for an agreed 5-day review, by which time the prospect is to call the presenter with feedback.

3. *Review the business plan.* The prospect reviews the business plan, calls the presenter to ask some questions midway during the review, and then calls to schedule another meeting.

4. *Meet to clarify the business plan.* The prospect asks the presenter innumerable questions, and continuing interest leads to scheduling of meetings with other players (casual meetings may have occurred already).

5. *Meet other players.* The presenter and the prospect meet individually, in small groups, or collectively with the other management team players, activities spread perhaps over days or weeks.

6. *Tour the project.* Intermingled with meetings may be demonstrations of various project aspects, plus informal meetings with key people distinct from the management team.

7. *Review linked aspects.* It also may be important for the prospect to review selected suppliers, consultants, subcontractors, or other entities critical to the project.

8. *Close the deal.* Finally, it is time for the prospect to invest or to depart.

You certainly do not want to play all your cards at first. Of course, you must reveal a lot to gain proper interest, but deeper project secrets should be progressively shared only under strict nondisclosure agreement control as a candidate successively shows more interest and commitment. There may be a number of elements that are not revealed at all until after the prospect has consummated investment. However, by law, you must provide full disclosure of all information relevant to an investment decision.

Throughout the process, the company and the candidate can increasingly examine each other, an essential step for mutual comfort and trust. An exemplary set of specific criteria to flag winning candidates follows:

- *Capacity.* The prospect must be able to invest an amount meaningful to the project, without personal strain.

- *Desire.* It is essential to verify that the candidate seeks actual investment, since some people explore projects for other reasons unproductive to the company.

- *Horizon.* Make sure the candidate has a proper investment horizon, realizing full well that dividends are unlikely for many years, if at all, and that there is not an established market for reselling company stock.

- *Expectations.* Related to the horizon criterion, verify that the prospect has realistic expectations regarding product and company development, stock share and its value, product, and short- and long-term company objectives and strategies.

- *Attitude.* The company must work with this person or group, so evaluate potential interactions during both good and bad times.

- *Participation.* Assure mutual understanding of the prospect's role in company activities relative to the following dimensions: active versus passive role in ongoing affairs, board member or not, mutually agreed nature and level of feedback, and mutually agreed method of resolution if disagreements arise and specified recourse.

Judgment and flexibility are required in evaluating and qualifying potential investors, because every person or group is different and special circumstances may apply.

7.4.3 Dealing With Objections

Even with the best candidates, objections or perceived obstacles may arise. You certainly want to anticipate as many of these as possible and develop sound responses to imaginable obstacles. Still, new obstacles inevitably will arise in the process of courting prospective investors.

Learn to handle such problems professionally. First, be sure to carefully understand the real issue. Question the candidate fully, to get beyond superficial complaints to the real root of concern. Avoid panic—the entire issue need not be resolved on the spot. Issues usually can be managed in many ways, and the real solution often is a combination of factors. One important response to an objection is listing other partially offsetting factors. There may be product design or marketing plan modifications or other program adjustments to lessen impact, or maybe research is warranted to help decide the best answer. Finally, we live in an imperfect world, where not everything is utopian. Objectionable aspects may exist that simply must be tolerated.

Those approaches—and surely more—can be employed to deal with any one objection. In any case, proceed slowly. Respect the objection, because it is real. Contend with it calmly and professionally, balancing considerations to best tackle the problem at hand. Handle multiple objections in the same way, combining individual concerns if appropriate but otherwise treating each distinct objection as if it stood alone.

7.4.4 Calling for a Close

Investors are people and so may have some quirks. One oddity often encountered is the candidate who just never seems to commit. You orchestrate one presentation stage after another, and the prospect seems to progress nicely. You have meticulously examined the candidate, and you feel the person would be good for the project. Everything is fine, except the prospect never commits to the offering as presented in the business plan and repeatedly reiterated in verbal discourse.

This situation resembles the one in which you have to nudge the review process along, except here you are courting a perceived winner. Take care to ensure sustained review process advancement, but when the arranged progression of review scenarios has been exhausted, it is time to call for the close. Ask if there is any further need for the business plan, because, if not, it is needed for other candidates. This effectively puts the candidate on the spot, diplomatically implying it is time to decide.

Some people continue collecting information for as long as it flows. When your presentation has run its course, ask the candidate to articulate any outstanding questions. If the troubles cannot be identified, there probably is no real interest. Alternatively, if real issues can be extracted, apply the previously discussed methods to deal with objections. Forcing a commitment decision may result in funding or a parting of the ways. Regardless, you must effect action and move on.

7.4.5 The Closing

Always incorporate before the initial funding search. Once an investment commitment is received, the individual investor or other legal entity must join the corporation, formally signing a written stock purchase agreement with the company. You and perhaps other team members are the existing corporation stockholders, and the agreement adds one or more new stockholders in exchange for stock. The agreement should carefully accommodate all disclosure and other requirements dictated by the private investment laws relevant in your locale. That ensures compliance with applicable securities laws and protects team members and the company. Your business attorney can help with the details.

The agreement may call for the prospect to buy all agreed stock at once, or installments may be scheduled based on time, need, or company performance. Any of those approaches is fine, but if the money comes in stages, then issue stock the same way. Never issue all stock for only partial payment; numerous circumstances and attitudes could change, potentially leading to damaging outcomes.

An important auxiliary consideration is the strategic investor, one who adds company capability or credibility. That is desirable, given acceptable adherence to all other previously mentioned investor criteria. Also, a large investment or significant capability or credibility enhancement often is given consideration relative to membership on the company's board of directors.

7.5 Ongoing Stockholder Relations

Once a candidate becomes a stockholder, the job is far from over. It is critical to keep each stockholder informed on company progress, extending funding candidate courtship. However, the process is even more critical here, the new stockholders now being part of the project. While not involved in daily management, new stockholders do help elect the board of directors, who, in turn, run the company (the officers report to the board). Besides, you do not want discontented stockholders; they have rights and can cause serious trouble if crossed. Finally, you may be able to acquire even more investment from friendly stockholders in later rounds of funding.

Several things help keep stockholders happy. Good company performance is one. Another is keeping them informed through reports on a regular basis. Quarterly periods are a good choice, although other intervals may do. Also effective is getting stockholders to participate in major decisions, one approach being to call a special stockholders' meeting to address the issue at hand. Finally, take time for some one-on-one interaction with each stockholder from time to time, to develop a rapport and acquire informal advice. These actions build stockholder attachment and allegiance to the operation, always a good management technique that can pay handsome dividends in a crunch.

By law, the company must conduct certain types and frequencies of reporting and must minimally hold an annual stockholders' meeting with certain minimal notice. The extras you do will cultivate stronger and more personal relationships. Privately funded company ownership usually comprises a small group of management and stockholders. The little things are eminently important.

7.6   Adjustments to Funding and Company Plans

As discussed in Chapter 6, plans change from time to time. It is senseless to adhere blindly to prior planning in the face of changing circumstance. That is why planning must be ongoing, constant attention being necessary to detect incessant change. Anticipating and contending with change are planning's charge, to provide, analyze, and interpret relevant data to define and, as necessary, modify company action. That applies not only to funding but to all company functions at all times.

The following examples present several forms of commonly encountered change and how prior planning might be responsively modified.

- *Funding.* Funding formally committed by agreement still may not materialize when scheduled, due to sudden investor-based problems. Planning must devise ways to secure alternative funding or to reschedule and perhaps alter activity priorities.

- *Management.* Say that a key management team member is lost without warning. Planning must effect some reallocation of responsibility among the existing management team, quickly find a replacement, hire an appropriate consultant, or alter activities and schedules to accommodate reduced C^3.

- *Product.* What if an essential component becomes unavailable just when it is desperately needed? Planning must pick among alternative part sources, temporary fixes, permanent redesign with concomitant delays, and the like.

- *Marketing.* Imagine that a carefully orchestrated exhibit at a major meeting suffers an 11th-hour unavailability of the centerpiece prototype machine. Planning must coordinate drastic marketing presentation changes, altered administration preparation for press releases and industry executive interactions, and reorganized technology division support of marketing and privately scheduled demonstrations.

All those responses by planning necessarily involve major extra efforts by other key company functions, because planning must acquire substantial data and interdisciplinary interpretation to arrive at rational new plans.

7.7 A Look Ahead

Part II discussed the primal flash, the original concept that seeds an entire project, and how it becomes molded and remolded many times into concepts for a product and an associated business. Those chapters examined planning, the business plan, and funding, three pivotal functions at the very heart of business development. As we end Part II, we have built a strong, deep and insightful background as a base for understanding business development. Part III goes on to develop the superstructure.

Actually, we touched on a small part of the superstructure in Chapter 4, which examined the concept, taken to mean the original project idea and its evolved form of product and business concepts. Much, if not all, of the latter development was said to occur in the concept stage of business development, the first stage of formal project activity.

Our goal now is to investigate the rest of overall business development, those several stages that occur after the concept stage. While remnants of concept development may persist awhile as maturing product and business concepts, their identities progressively dissolve into postconcept embodiments such as actual product development and design and formal business structure. Part III, Product Design and Launch, presents a comprehensive treatment of central business development stages and the rationale for their design, content, and management.

Part III
Product Design and Launch

To this point, we have traced the creation and development of both product and business concepts and built a broad base of facts and methods regarding planning, the business plan, and funding. Now, in Part III, we delve deeply into the central stages of business development. Chapters 8 through 11 describe the four consecutive stages of business development that follow the concept stage (Chapter 4). The stages start with the earliest research and extend through product launch and initial sales. Recall that stages are named according to the most visible activities during the given period. Nevertheless, discussion carefully encompasses the primary contributions by all key company functions at every step.

The R&D division is charged with bringing the project to full readiness for the stage to follow, the design stage. That means the technical division of the company must create, evaluate, compare, select, and formally verify a well-defined design approach. The administrative division must ensure that all resources are on hand when needed and that overall planning and the business plan are current and complete. Finally, the marketing division must design, acquire, and interpret all background data for the major marketing plan activities of the next stage.

The flagship of the design stage is detailed product design, and the technology division must begin to formulate the particulars of product manufacturing and of QC activities as well. In parallel, the administrative division must do its usual thing: supply resources, conduct planning, update the business plan, and acquire funding. The marketing division must fully develop the formal marketing and sales plans and embark on early ideas and development toward implementing those plans.

The launch preparation stage follows the design stage. Administrative activities are not much different in kind, just more advanced. The technology division has completed product detailed design, and manufacturing, quality, and product shakedown are now the central issues. Marketing is focused on implementation and product launch.

Finally, attention turns to the product introduction stage. The technology division is feverishly trying to squeeze every bug out of product detailed design and manufacturing, and QC is busy installing procedures to monitor and correct problems. Administration must contend with mounting resource demands, and planning, the business plan, and funding needs are arguably the most demanding and dynamic of any period. Chief concerns of marketing include final posturing efforts, the advent of promotion, product launch, and the startup of sales and allied activities. During the product introduction stage, the company's future rests heaviest on marketing.

8

Research and Development Stage

8.1 The Path From Concept to Design

A sketch of an idea precedes a developed concept, and concepts predate designs, which come before, successively, production, marketing, sales, and (one hopes) success. Outlining the series is simple, but it is a colossal job to carry it out.

The one business development interval that occurs prior to the subject of this discussion is the concept stage (see Chapter 4). The most visible company division during the R&D stage is the technology division. Here we take a fresh look at the two closely allied technology functions, research and development.

- *Research.* This function assesses product idea feasibility and the best approach to a workable system. Research combines both literary and experimental activities to ascertain if the product idea is feasible.
- *Development.* This function extends the analysis to determine if the product is feasible and if it is, to define a specific design approach. Development must ensure a product with designated features and acceptable development and production costs, applicable to defined markets under suitable marketing cost/benefit assumptions.

Despite the visibility of technology, the R&D stage must advance the entire complement of product and company objectives. In that view, all three company divisions, administration, technology, and marketing, are addressed below.

Technology takes center stage in R&D because product development is the organization's early driving force. But administration and marketing are indispensable, too. Administration provides the environment within which the project exists. Administration originally forms the project, does the business planning, creates the business plan, and provides facilities (physical or virtual) where project activities occur. In turn, marketing deeply influences both other divisions, through user-oriented inputs that significantly shape product development (technology) and the knowledge, data, and wisdom that strongly influence planning, the business plan, and funding (administration).

This book shows how to turn a product idea into a successful business. To reduce complexity, simplifying assumptions have been made. One is that all projects fit the exact mold of this story. Of course, every project is different, some fitting better than others. In any case, you should feel free to adapt book content, if circumstances warrant. The story we tell is internally consistent and complete and usually will work basically as is for a project without extraordinary elements. One of our goals is to convey the wisdom to recognize when such adjustments would be best for your project.

Until now, we have imagined project elements strictly as concepts. There has been no manipulation of anything tangible, short of primordial paper or computer design conceptions. Your project might be different. Prior to the R&D stage, you may have wanted or needed to actually test an imagined mechanism or other product aspect. Some issues could be so compelling or desirable that even early concept feasibility could not be properly assessed without such tests. We propose a prototypical sequence that optimizes resources, barring extenuating circumstances.

8.2 The Product Research Process

We assume the input to the research stage is written product and business ideas summarizing all concept stage efforts. The notions are written to carefully register the compounded results of feasibility analyses conducted in the concept stage (see Chapter 4), the latest embodiment of product design records (that we recommend be kept for future development reference), and to document design decisions in case of litigation.

Research relevant to the privately funded project tier of complexity is segregated into several categories. Within each group, we list a number of key topics for investigation. Every item may not apply to all projects.

8.2.1 Literature Research

We have mentioned literature research before. But the project is now more advanced, and the central questions are either different in nature or degree.

We live in the information age, buried in data, but most budding entrepreneurs pass up volumes of valuable information that is readily or reasonably available. Do not reinvent the wheel—check the literature. For one, the reinvented element may not capture full functionality. Regardless, it is a waste of precious resources to recreate something that already exists.

Literature research can provide significant support in the following product and company functions.

8.2.1.1 Competitive Comparisons

Two comparisons are essential. The first is a comparison of all competitive products currently offered in your marketplace. The second is all products you suspect might arise from existing or new competitors. The second comparison is tougher. Handle it by listing every conceivable product improvement or spin-off of company and competitive products. Develop a technology profile of each competitor by analyzing past and present designs. This may reveal patterns suggesting who will do what next.

If you are successful, the competition will respond, so garner as much product protection as you can (see Chapter 6). Also, differentiate your product through promotion, service, sales, quality, and any other way possible. Finally, when the time comes, enter the marketplace quickly, to establish an early beachhead prior to competitive responses.

8.2.1.2 Financial Modeling

Conduct a comprehensive reanalysis of the 5-year financial model, to create an updated estimate of the timing and the extent of every product and company activity occurring within the 5-year window. Model updating can require significant research at any business development stage, but successive modeling rounds should provide progressively harder data on revenue and costs as the project advances. Literature research is important to optimize design and production approaches and details and to begin organizing thoughts regarding product marketing.

8.2.1.3 Financial Feasibility

Management should reassess financial feasibility every time the model is updated. The reliability of project feasibility evaluations rests on the quality of model assumptions and data, in light of predicted funding requirements and investor return. Identify major project risks and weaknesses, reduce each risk as much as possible through creative planning (see Chapter 5), and adopt realistic operations and funding plans to optimally contend with remaining risk. Freely use the venture capitalist thought experiment (Chapter 4) and selected outside experts to test your approaches.

8.2.1.4   Downside Risk

Another useful risk analysis is to estimate the likelihood and financial effects of project failure at defined times. General business development books can provide examples of failure modes and causal interpretations. Upon failure, the loss could equal all invested money, less applicable assets. Or losses might exceed investment, if harm could pierce the corporate veil and assault personal assets of the main players. The loss estimate should ignore time and other donated resources (deal only with real money). Finally, try to lower losses and risk through creative planning.

8.2.1.5   Upside Potential

This is the flip side of downside risk. Gauge what the company could bring the key players under various successful scenarios, derived directly from model versions run under different assumptions. In addition, judge the probability of each scenario and see if prospects and rewards can be improved through judicious planning.

8.2.1.6   Legality and Safety Issues

This catchall class helps ensure that planning has designed the most effective approach and fully covered all associated costs and resources. Relevant topics include patent, trademark, and copyright protection for company assets and assurance that designs, designations, and production techniques do not infringe on protected others. Also, be sure that all applicable laws, regulations, standards, and guidelines are properly addressed, including less obvious entities such as the Occupational Safety and Health Administration (OSHA) and consumer groups. Finally, consider proper labeling, legal support, and insurance. Key literature sources include patent and business books and magazines, industry and trade associations, small business development centers (usually state sponsored), and relevant government agencies.

8.2.1.7   Market Issues

The R&D stage involves commitments to a specific product design approach. It is imperative to evaluate the market size (all conceivable customers) and market (expected sales/year) once again (see Chapter 4), but more stringently. Prior estimates were preliminary, but impending technical and marketing committals render a current analysis crucial. Expand your marketing research, bring every team talent to bear, and seek outside help if contacts or resources permit.

8.2.1.8   Sales Forecasts

A critical addition to the market considerations is forecasting sales, a tough but necessary process to support planning and business plan refinement. The marketing research people should review relevant trade and educational literature, conduct surveys, acquire commercial market segment studies if available, talk to potential customers and experts, hold focus groups, attend trade and technical meetings, and do everything else of reconnaissance value. They then should build a detailed financial sales model based on a realistic interpretation of the data. Actually, multiple forecasts can be useful, representing relatively high, medium, and low sales. If sales are highly uncertain, all three forecasts might be placed in the business plan, but the business plan usually will include only the scenario considered the most realistic.

8.2.1.9   Marketing and Sales Costs

Sales estimates must be based on a clear picture of the total marketing and sales resources required to generate them. Crudely, a high sales forecast usually is incongruent with low marketing and sales costs. Marketing must lay out in detail exactly what marketing activities will be required and their cost; the same must be done for sales activities. It is best to project what is believed most realistic, on a scale spanning conservative to optimistic extremes. If possible, add a contingency for unexpected marketing and sales costs to help protect the sales numbers, since the whole project depends on sales. Useful literature sources include consultants, research firm industry studies, trade associations, small business development centers, and published government data.

8.2.1.10   Product Design

Product design certainly affects sales. Sales prowess is a direct function of product functional utility, all else being constant. The R&D stage defines the final vision of the product, prior to detailed engineering design, so here the charge is a literary product shakedown using, for instance, competitive product literature; trade magazine articles; various industry meetings, courses, and forums; and consumer watchdog literature. Carefully analyze the design of competitive products. Examine how specific features interrelate with other features and overall product design (note impressive aspects of their design) and analyze the competitive design relative to materials requirements, production costs, ease of use, and the type of manuals or labeling included. Carefully inspect promotional material and actual product for signs of regulatory, safety, and standards compliance; identify what is unique about the competitor's packaging and shipping; and so forth. Usually, special issues will be referenced through labeling or promotional material by requirement or to convey producer concern and attention.

8.2.1.11   Manufacturing

One focus of the research stage is a final check on manufacturing before development. Prior assessments are extended to look deeply at how the product concept could best be produced. Search for special needs relative to materials, processes, methods, tooling, or setup requirements that might dictate high cost or other obstacles. Also examine forecasts of sales relative to factory throughput and space needs for operations management, purchasing, parts, assembly, testing, rework, packaging, finished product inventory, and shipping and receiving. Ensure appropriate allowances for quality and manufacturing control, including personnel and equipment. All these considerations are relevant, whether you select in-house or contract manufacturing, because they all affect cost and quality. However, your selection will greatly affect the details.

8.2.1.12 Talent

Essential to planning at all stages is ensuring proper staffing. The typical product-oriented, privately funded project may go for quite some time with only the management team on board, but the R&D stage drives many projects to take on other help. Regardless, it is up to planning to dutifully add people as needs and resources permit. One indispensable attribute required of early company players may be multiple talents. At first, such people can offer help in several areas, before assuming more consolidated roles with further growth. When hiring, consider potential roles from both the short-term and long-term perspective.

8.2.1.13 Administrative Issues

As usual, the chief concerns of administration relate to general management, planning, the business plan, acquiring and maintaining the needed facilities and resources, and, of course, securing funding. It is often the R&D stage or its successor, the design stage, in which major people, equipment, parts, fabrication, and outside expenses first appear.

8.2.2 Contacts With Knowledgeable People

We now turn to people as a source of product and company evaluation. The same list of topics used for literature research could serve here as well. But we chose a distinct list of topics, to provide new perspective. In fact, your group can combine the two lists to assess both literature and people research from an even broader viewpoint. The various items were selected to emphasize the use of outside expertise, but significant in-house scrutiny should be applied as well.

8.2.2.1 Needs and Desires

One parameter benefiting from outside scrutiny is how well the product meets perceived needs and desires. Real product needs tend to foster product demand even in tough economic times (e.g., computer supplies, software and equipment that dramatically increase productivity, and so on), while nonessentials do not fair as well. Objectively assess where your product stands in the relative pull by customers for essentials versus the marketing push needed to create noncritical sales. Important considerations include pricing, perceived utility and value, appearance, nature of the customer, distribution channels, and trends in needs and desires over time. Useful sources include consultants, industry and academic experts, and small business development center personnel.

8.2.2.2 Buyer Description

A profile of typical buyer categories is usually critical. Combine in-house analysis with pertinent outside help to identify buyer groups from marketing research data. Then form a profile for each type of buyer and design marketing and sales plans to specifically target or at least exploit the profiles. Design future marketing research around the profiles, to further dissect and characterize your buyers. Key contacts include consultants, industry experts, product users and buyers, and customer-based or commercial maintenance groups that maintain and service hardware and software systems.

8.2.2.3 Marketing, Sales, and Distribution Plan Viability

Outside help to assess the viability of the marketing, sales, and distribution plans can be pivotal. Objectivity is difficult but especially crucial for issues as vital as marketing, sales, and distribution, since the company is likely quite dependent on correct assumptions and on plans that really work.

8.2.2.4 Customer Service Plan

Be sure to review the product from the buyer's standpoint. Actual users of competitive products are an excellent source of data, but, as always, any expertise is welcome. The point is to ensure that general customer expectations regarding service, maintenance, warranty, product information, and any other questions or problems can be met acceptably. The best approach is to see how your particular industry generally performs in that regard and then assess how you might need to be different. Some market entry strategies focus on superior customer service.

8.2.2.5 Economic and Political Factors

Ensure that your plans allow for seasonal variations and other cyclic or irregular changes in given markets. Sources of data include government agencies relevant to your industry, trade or technical associations, and industry analysts. Similarly, seek out people knowledgeable of political factors like upcoming laws and regulations that may change the way a market works, such as new transportation, production, design, or promotional regulations.

8.2.2.6 Product Viability

Outside sources are especially useful to help check the true functional utility (efficacy) of a product (it is easy to fool yourself due to emotional attachment). Other related concerns are product price, safety, reliability, quality, durability, and perceived value relative to competing alternatives. Technical and marketing experts and customers of competing products are particularly useful here.

8.2.2.7 Business Plan Viability

Another superb place to elicit outside input is on the business plan. Expert advice is best (academicians, business plan critique groups, successful entrepreneurs and corporate executives, accounting and legal firms), but much can be learned from the casual bystander as well. Use every type of available source for review and comments, and weigh feedback value accordingly. All or the majority of initial project knowledge for most investors will be limited to the business plan. Only after reviewing the plan will they entertain direct dialogue, a situation perhaps frustrating for you, but it saves the prospects time and effort. Key attributes for business plan evaluation are to ensure that all major issues are covered (see Chapter 6 and Appendix A), the presentation is clear and well written, and the offer is clearly understood and attractive.

8.2.2.8 Design Approach

During the research component of the R&D stage, it is time for early thoughts of product design, specifically the general design approach. Assume the product is a powered golf cart. We might elect conventional electric or gasoline power, like everyone else, or we could choose nuclear power to provide differentiation. Defining the design approach is frequently monumental, since everything may flow from this pivotal choice. By choosing a nuclear-powered golf cart, we pursue a specialized market distinct from traditional products. We also commit to unique requirements for design and manufacturing and enter a stringent regulatory environment.

8.2.2.9 Future Related Products

One-product companies have problems with funding, due to inherent vulnerability. First, a sudden competitive move can be disastrous, there being nothing on which to fall back. Second, the firm is less protected against economic and political perturbations. Third, customers are fickle, and if preferences, attitudes, or traditions change, they may wander off. The best defenses against such happenings are diversity (many avenues), planning (early warnings), and adaptation (corrective responses). Plan with ultimate product diversity in mind, cross-check your plans with outside input, and give this dimension a lot of weight (investors do). Both breadth and depth are important. Breadth means multiple offerings in parallel, closely tied paths, such as different powers of desktop computers. Depth implies covering many functions and paths, as in producing many types of computer (palmtops, desktop PCs, notebook PCs, servers, workstations, and networked systems).

8.2.2.10 Life Cycle of the Product

Products have differing life cycles. Laptop (luggable) PCs were taken seriously for only a very short time, being displaced by notebook PCs shortly thereafter. Examine your product relative to life cycle, the possible ranges spanning short-term fads to products satisfying long-term needs without obvious alternatives. A business with a good mix of long-term products typically is more attractive, but special circumstances might render other scenarios feasible. The most important consideration is realistically defending whatever planned product mix is adopted.

8.2.2.11 Impact on the Environment

This issue faces rising modern concern in many, mostly advanced, countries. Key factors include whether packaging can be recycled or degraded; whether component elements are hazardous to land, water, or air quality; the use of such items in manufacture; consumption of hazardous components during product use; and ultimate disposal of expended product.

8.2.2.12 Impact on Society

Finally, consider the product's impact on society. Pursue outside opinions if this attribute is questionable or looms large. One subject is compatibility with current society and its norms, traditions, and practices. Another is the relative good or bad nature of the product, relative to safety, lifestyle, culture, and prevalent belief systems. Computers generally are regarded as benefiting humankind, but weapons have a mixed record.

8.2.3 Tests of the Concept

Some form of subsystem testing or other tangible evaluation may occur before the R&D stage, depending on the product. However, testing is even more probable during R&D, since a specific design approach must be selected prior to stage completion. R&D must serve to catch bad design approaches as well as identify a chosen one, since a later need to backtrack during detailed design to create a new design approach would be prohibitive in time, money, emotional energy, and credibility.

For simplicity, testing can be reduced to two classes. One is analytical, in which paper and pencil (or the computer equivalent) are the main tools of action. The other class is experimental, in which some form of hardware or process is set up and run through a test protocol.

8.2.3.1 Analytical Feasibility Tests

Analytical testing involves subjecting the product or selected subcomponents to simulated testing. Simple tests may involve manual calculations to verify fit and function. One illustration would be to ensure that a new car wheel design actually conforms to the size(s) of tire targeted for application. Another instance would be verifying a new chemical process using small laboratory quantities. Before the days of PCs and workstations, complicated procedures such as finite element analysis had to be done by hand, the approach usually involving drastic simplifying assumptions to render calculations feasible. Of course, the danger was that oversimplification could render the model unrepresentative. Simplifying assumptions are still employed, but advancing computer power has rendered even many complex problems subject to much more meaningful analysis.

Today we enjoy a wealth of computer hardware and software to aid analytical product or subsystem feasibility testing. Whole shelves of many technical libraries are filled with pertinent books on applications and techniques. Computer tools now exist for a myriad of analytical tasks, including capabilities of research stage feasibility testing and far beyond.

Simulation affords huge advantages to projects that can employ the described technology. Any such project should seek a management team member knowledgeable in this burgeoning arena and then lay out the project feasibility and design program to take maximum advantage of simulation and computer design. The enormous increases in efficiency and savings in time and money should pay handsome dividends.

Many products are not suitable for computer simulation, but simulation merely exemplifies extensions of basic computerized design, which is virtually mandatory today. Always use the best evaluation and design technology applicable in your situation, to optimize productivity and to facilitate design documentation transferability between independent groups, both within and outside the company. Rarely can anything beat computerized design, the most popular example being Autocad®. Even future utility is a strong argument favoring computerization, since correcting a design or engineering the next generation system often is far easier when the documentation is computerized.

8.2.3.2 Experimental Feasibility Tests

Some issues of feasibility require tests conducted on real hardware or using genuine processes. Tests involving actual hardware are critical for evaluating such items as permanent medical implants or high-vacuum seals. Regarding processes, chemical engineers routinely test feasibility in small laboratory setups, carefully designed for scaleup to full capacity later. Another process example involves functionally verifying complex automatic DNA sequencing systems.

Experimental feasibility tests imply running one or more trials of a physical system under a defined protocol to collect feasibility data on the system or some subset thereof. Simulation should be used in situations so clear that real system performance is unquestioned. However, if simulation is not possible, practical or unambiguous real-world testing is necessary.

Let's say that an experimental test is deemed essential. The results of any experiment are only as good as the experimental design. We have seen many people fooled by experimental results, because the experiments did not really answer the questions being asked. Poor experimental design is a waste of time and money, and the results can be dangerously misleading. Isolate one or two questions for study and ensure that the test protocol addresses the target questions in a fashion in which all external variables of influence are eliminated or controlled. It is the latter point that causes the most trouble, there often being subtle extraneous factors that can affect the results. Control experiments may be required to quantify those intrusive factors, if protocols to exclude them are impossible or impractical.

Say we want to test the performance of an exotic new computer chip. Of course, we must subject the chip to every conceivable class of processing, to verify first-order electronic performance. However, for said data to be useful, we also must duplicate normal operating conditions pertinent to real-world chip performance, such as electromagnetic interference, temperature, humidity, and perhaps other relevant influences of nearby devices, components, or other environmental factors. Failure to chart chip performance throughout all meaningful permutations of the expected range of operating environments renders the data suspect and therefore of little value.

This book targets people with a technical background, and many already may have a good feel for the preceding points. Just be attentive to the possible missteps. Carefully review all protocols in meticulous detail and avail all the outside inputs you can muster, expert or otherwise. Ensure that reviews occur with data distortion firmly in mind. In short, protect yourself as much as possible from false conclusions, since undiscovered design flaws can lead to expensive redesign, product recalls, and even total customer disenchantment.

8.3 The Product Development Process

Having completed the research component of R&D, the baton now passes to development. At this stage, the technology division carries its analysis a step further, to decide if the product is feasible and to define a specific design approach. In turn, the administration division further analyzes the business idea to determine if a solid business is possible, defines its nature, and sets the base for future stages. Finally, the marketing division helps with product development, planning, and funding and further develops background for later marketing and sales activities.

Completion of research implies product idea viability. There is a big difference, though, between a viable product idea and a viable product. The subheadings and discussions in this section highlight key elements required to bridge the gap. Discourse is limited to the technology division of the company; the administrative and marketing divisions are treated later.

8.3.1 Product Definition

It is important to lay out the general boundary conditions of the product. Many will state that the nature of the product has been known all along, but experience shows that what people think is well defined mentally suddenly becomes elusive when they are asked for details. Some do not realize that the product was not well formed until they are pressed, while others do not feel that such questions are fair until the design stage. Both reactions are wrong. You must begin to shape and bind the product early, so the good and bad of various approaches can be thoroughly evaluated.

8.3.2 Design Approach

Development represents the last chance to check and adjust the design approach. First, isolate and identify functional subelements within the overall system and define specific mechanisms by which each subelement will function. Subsequently, use the same complement of evaluation methods developed in Section 8.2 (literary research, expert inputs, hardware and/or software experiments) to test critical aspects of candidate design approaches and to select between multiple approach choices. The goal is to explain and defend every aspect of the design approach, noting that this is not yet the design stage. For instance, a design approach might read "control of temperature will be effected electronically, using data from an array of sensors." You would encroach on detailed design (the subject of Chapter 9) if you actually specified the sensor type or details of circuit design.

8.3.3 Functional Utility

Functional utility is about the real purpose of the product, what it is supposed to do, and how. A final review of the nature, size, functions, and other user-related attributes is crucial. Every available analytical tool should be brought to bear to ensure the intended product serves user operational needs in the best way practical. The three analytical classes described in Section 8.2 (literary research, expert inputs, hardware and/or software experiments) cover the spectrum of available evaluation tools.

An illustration may be helpful. Consider a flow soldering machine that has sound adjustments for all pertinent variables (e.g., part depth and orientation, solder temperature and composition), but that does not include any kind of diagnostic capabilities to specify the effect of every variable on finished product. Electronic circuit boards might exit the machine with solder joints excellent on one left-to-right aspect but poor on the other. Without a simple and quick diagnostic system available, operators might spend considerable time discovering a slight board tilt, temperature anomaly, or other problem. The functional utility of the flow solder machine design is weakened in the absence of practical diagnostics.

8.3.4 Manufacturing and Shipping

It is important at this point to thoroughly examine how the product is expected to be made and transported. Postsale activities such as product setup and operator training may be relevant for some products as well. Early thought is critical to reveal any particularly critical problems that must receive concentrated attention throughout design due to special impact.

It is true the results of detailed design are not known yet, so considerations of manufacture and transport may seem premature. Nevertheless, much is known about the general nature and attributes of the product, as we see from the other development activities. Development, like all other stages of business development, is rife with recurrent thought and action. Issues may be addressed at a general or superficial level early in the process, and then the same topics may be revisited many times as the design advances. The business development trajectory is better viewed as a spiral of progress rather than a circle of repetition.

8.3.5 Cost Versus Price

The real topic here is profitability. Profitability commands high attention at every stage of planning, but the focus now is development, the last chance to ensure decent margins prior to the design stage. Every aspect of cost must be examined, including parts and labor expenses, testing, packaging, shipping, setup, training, and every other element of functional delivery to the ultimate user. In addition, compare estimated total product development costs to expected sales over the product's lifetime and evaluate product viability in that light. Recheck all data and reasoning behind sales figures and assumed pricing, because the entire company likely depends heavily on early product success.

8.3.6   Quality and Reliability

Quality and reliability must be designed into the product, being inherent in the deepest elements of design. In other words, quality and reliability must be integral to the design of individual components and the way the components interact. Advantages include lower long-term design process costs and a smoother running shop with happier management, employees, vendors, and customers. Only the foolish try to instill quality and reliability during final test by reworking deviant product.

8.3.7   Maintainability

Ultimate product maintainability is frequently critical, although easy to sidestep, because its impact is remote and so many other issues need attention. Even so, long-term success is best served by nurturing favorable feedback from all customers, a process beginning at the advent of product development and extending forever. One key is offering products that are easy to keep in service, requiring sound design in three distinct areas: reliability and ease of use; maintenance and repair that are easy to understand and conduct; and clear and complete documentation of product use, repair, and maintenance. Apply these critical notions at every stage in the product development process, since this is the last chance to instill maintainability in the product prior to the design stage.

8.4   Technology Transfer From R&D to Design

Many elements defined in the research process get modified and refined again during development, while other elements are relatively unchanged. Regardless, technology transfer must occur to convey R&D technology to the next development interval, the design stage.

Here we analyze the "how" of R&D from a fresh vantage point to enhance perspective, the magic ingredient first introduced in Chapter 2. After completing technology transfer, we close the chapter with key R&D stage activities of the administrative and marketing divisions of the company.

Technology transfer is important, because different people may conduct R&D versus detailed design. In any case, like any other act, documentation of design approach decisions and rationale is critical as a record for future reference. This is necessary because people may drift if not kept on track by other people and past decisions. Old conclusions can be easily clouded by current pressures and new ideas.

The following list sets out key elements in technology transfer from R&D to the design stage. The list is not exhaustive, but it does suggest useful ways to ensure continuity of thought:

- *Preliminary detailed design.* Initial elements of actual detailed design are common during development or even earlier, because it is often necessary to consider some future particulars in the process of preliminary development. Still, ideas at this stage are likely to lack coordination and specifics, and ideas for distinct system elements may even involve inconsistencies, if such conflicting relations are believed subject to resolution later in design.

- *Preliminary testing.* As early designs of certain elements appear and interact, testing becomes necessary for ultimate selection. Here, testing is viewed broadly, to cover all types of literary, expert, and experimental appraisal (Section 8.2).

- *Prototyping.* At key points of progress, it often becomes essential to assemble some form of a given subcomponent or the entire system for key tests. Such prototypes can be of many forms, from a nonfunctional rendition for form and fit data, to bench tests that show function without fit, to full working prototypes with some mix of either prototyped or production elements.

- *Iteration.* It is common to see design, testing, and prototyping activities that spiral into advanced rounds of the same sequence. Many spiral cycles may occur as the design evolves.

- *Marketing spec.* In one view, the final act of development is to produce a marketing specification (marketing spec), a product description cast in user language. We adopt an alternative view, in which the marketing spec represents the first act of the design stage.

The list is a sample outline; a unique program must be adopted for your group consistent with product and company.

8.5 Administration During the R&D Stage

Much of the discussion in this chapter has involved the company's technology division, but the administration and marketing divisions play key roles, too. Here we discuss administration, marketing being treated in the next section.

Administration's main thrust is to provide the necessary facilities, people, nonhuman resources, planning, and funding for the other two divisions, technology and marketing, to do their jobs smoothly.

It may seem that administration is in a rut where nothing changes, but that is illusory. While administrative task designations are similar over time, the challenges change substantially. During the R&D stage, the primary administrative priorities include:

- *Facilities.* The need for space can be substantial, at least for the technology division, in spite of likely cash limitations. Therefore, try to base company activities in the homes of management team members as much as possible, to minimize the need for central facilities as long as possible.

- *Human resources.* As R&D intensity rises, it is usually necessary to hire extra technical people, to handle necessary R&D activities. In turn, it is best to control the use of administrative and marketing people, beyond the core management team, to preserve cash.

- *Nonhuman resources.* Some projects demand tremendous equipment, parts, materials, services, and other factors, while others need little. Whatever the need, the watchword is frugality, combined with ensuring that every essential element is fully addressed.

- *Planning.* Here lies the linchpin for all else. Proper planning cross-links complex activities into a network of parallel yet interlaced processes, creating coordinated advancement and mutual support. The secret at the R&D stage is to support current activity and to lay groundwork for the explosive stages to come.

- *Funding.* Every company function is critical. By the R&D stage, hardly any function can properly advance without some funding, but the nature of R&D funding is paradoxical. Suddenly, much funding may be needed, but the company has little tangible material to show investors. Planning and funding must together build a carefully choreographed professional show that exudes credibility. Excitement is crucial too, but investor excitement cannot be created through displays of team emotion. Design the program so the excitement builds naturally from project and presentation content.

These central concerns of the administrative division are pivotal. All long-term project goals depend on successful administrative performance.

8.6 Marketing During the R&D Stage

We have examined the activities of both the administrative and technology divisions during the R&D stage. Here we list the concerns of the marketing division.

- *Marketing research.* The literary and expert aspects of the research stage discussed in Section 8.2 require significant help from marketing. Key issues like early customer descriptions, user needs, and relevant markets cannot be totally left to technology people, who lack proper methodology, knowledge, experience, and perspective.

- *Identify market(s).* A prime R&D marketing task is identifying one or more relevant markets, now that the envisioned product is becoming more concrete, extending the prior list item that addressed marketing research. So R&D stage marketing research should be multifactorial, including two more factors in the next item.

- *Preliminary marketing and sales plans.* By the R&D stage, the marketers must begin preliminary plans for marketing and sales, at least a broad outline for action at later stages being formulated during R&D. That creates early structure and questions that can then be carefully fleshed out and addressed as the company advances. The same is true for all company functions. Plan well ahead. The alternative, rapidly forming plans on the fly just in time for action, is fraught with risk. Marketing topics of concern are the standard ones detailed elsewhere, customer characterization, program design, and product imaging, promotion, pricing, forecasting, distribution, sales, and customer service.

- *Product features.* Many engineers feel that marketers are just in the way, but their role is essential. We noticed that a long-time leading pool table maker came out with a new line some time ago. The tables were magnificent furniture but performed poorly, the design obviously lacking appropriate user-relevant advice and feedback. Make sure marketing is involved in product concepts and R&D from the very start and at every step. The marketers are not meant to displace engineers and designers, but they must be constantly in the loop to ensure realistic and functional design, one of their indispensable roles.

- *Contributions to planning and funding.* Marketing inputs are also obligatory with regard to planning and funding. Administrative and technology people cannot in any way substitute for marketing expertise. All three disciplines are critical and unique. Even initial planning, which arises long before R&D, must completely and professionally address marketing and sales. After all, marketing typically attracts every nickel the company will ever make. Also, potential investors want to hear about marketing and sales plans directly from the responsible parties, since third-party summaries will not do.

The R&D stage is witness to dramatic developments in all three company divisions, administration, technology, and marketing. The purpose is to prepare for the stage to follow, the design stage. Notice that the entire company must be ready for the next step, because no one function can carry the ball alone.

9

Design Stage

9.1 The Marketing Spec

So far, our discussion has progressed through the conceptualization and R&D stages of business development. The next interval, the design stage, is marked by three distinct specifications, or specs. This section examines the marketing spec; the design spec and the product spec are addressed in Sections 9.2 and 9.4, respectively.

The sequence of specs is logical, following the natural flow of product definition (Figure 9.1). The first step involves defining what is termed a marketing spec, a written description of the product and what it achieves, in the language of the user. Section 9.2 shows how translating the marketing spec into engineering terms leads directly to the design spec, the technical reference that design is intended to fulfill. Once detailed design is complete, a product spec can be coined, which defines in technical terms exactly what the final design accomplishes and how it works.

Recall that R&D stage product activities are mostly technical, but marketing plays an important supporting role, keeping the advancing product on track with targeted user needs. Still, the primary documentation that accompanies the product and records its evolving form is technical, a superficially ideal situation since no transformation of information from R&D to detailed design is required. However, recall the example in Chapter 8 that exemplifies a prime danger: The design target was a quality pool table, but the resulting product was nothing better than a large, expensive, and not very functional piece of furniture.

The output of R&D is a technical description framed by technical people in technical language. Without nontechnical input, however, the description may be incomplete and have inconsistent elements from the standpoint of

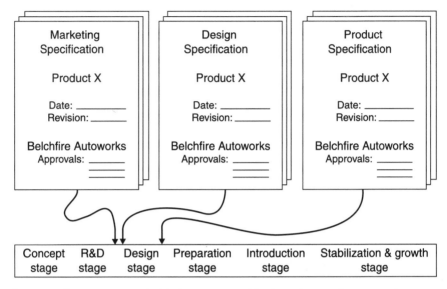

Figure 9.1 Three recommended design stage specifications. Arrows show approximate occurrence during the indicated business development timeline. Even though each spec originates at the indicated points in time, each spec may be revised from time to time as the product design process advances. Full approval by all key functions should be required at every spec revision stage, including spec origination.

the end user. Also, putting all the elements together from the R&D description would not necessarily result in a complete product. Gaps are likely, since detailed design has not yet occurred. The technical R&D description is often a collection of loosely bound subsystems or mechanisms envisioned to contribute to the ultimate product, but many subcomponents may lack meaningful coupling or coordination.

Of course, this situation easily could be remedied through a specific design effort focused on pulling everything together. In fact, this entire chapter is about just such an effort, the design stage. But here we address only the first detailed design step in the extended design process.

The marketing spec embodies the following critical interaction that we highly recommend for added design assurance. Before any engineering design, have the marketers lead creation of the marketing spec, forcing everyone, including the technology people, to develop a user-level description of exactly what the product will be and do. If our hapless pool table designers had properly conducted a marketing spec process, they likely would have avoided disaster.

Figure 9.2 includes an example of a marketing spec. Notice that the document does not detail how the product is to look or work. In pure form, the

Marketing Spec	Design Spec
1. Active heat	1. Electrical resistance material
2. Controllable intensity	2. Potentiometer control
3. Ease of interaction	3. All-purpose user interface
4. Portable	4. Battery power
5. Rechargeable	5. Rechargeable batteries
6. Low price	6. Cost < x (proprietary)
7. Heats for at least 3 hours	7. ≥ 3 hours, low heat setting
8. Machine washable	8. Mild, convenient conditions
9. Sheds rain	9. Waterproof exterior
10. Easy transport system	10. Bag for system components
11. Lightweight	11. System weight ≤ 8 lbs

Product Spec
1. Resistance wire (proprietary)
2. Pot regulation of duty cycle
3. Specific layout (proprietary)
4. ⎫ Specific rechargeable
5. ⎭ battery (proprietary)
6. x – $0.50 (exact amt propr)
7. 3.25 hours on low heat
8. Specific instr (proprietary)
9. Waterproof Oxford Nylon exterior
10. Specs for custom bag (proprietary)
11. System weight 7.5 lbs

Figure 9.2 Marketing, design, and product specs for the same product. The figure shows a one-to-one correspondence between specs to emphasize required direct correlation. Several product spec items are referenced generically here, to protect proprietary content (the subject product was taken from the real-world example in Appendix B). The items of each spec are illustrative only and are not intended to be complete.

marketing spec merely says what the product is expected to accomplish, not how. Even so, the real world is messy, and not all marketing specs can be totally pure. For example, the colors of medical device indicator lights relevant to certain functions (system energized, danger signal, battery charging, and so on) are defined by international standards, so the design group has little or no latitude. Product cost is another common boundary condition. The point is that the marketing spec places few or no unnecessary constraints on engineering.

Listed next are sample categories of marketing spec items for a generic electromechanical hardware product. The list is not intended to be exhaustive but is meant to illustrate important categories of consideration.

- *Dimensions and weight.* Configuration and modularity are additional aspects important for some products.

- *Aesthetics.* Appearance is almost always important, but some products dictate specialized industrial designers to create the level of aesthetics appropriate for the marketplace.

- *Primary functions.* This represents specification of the main actions or processes central to the product's intended use.

- *Secondary functions.* The product may also provide ancillary actions, processes, or capabilities not critical to primary function but useful relative to user convenience or productivity.

- *Ease of use.* Most products are meaningfully enhanced by due design attention to all genuine means of product use.

- *Portability.* Products of all sizes may benefit from considerations of device transport or more localized mobility, including attention to power requirements or other specialized access needs.

- *Quality and reliability.* It is critical to have a clear definition of required and desired quality and reliability prior to any detailed design.

- *Cost and price.* From preliminary data on intended markets, define a target customer price, which in turn will set limits on COGS (cost-to-price ratios of 1:3 to 1:5 are common, depending on the industry).

- *Documentation and training.* Carefully define descriptive, educational, and orientation needs of all customer, buyer, maintenance, and regulatory entities before conducting detailed design.

- *Cleanability and maintainability.* Easily overlooked are considerations of how the product can be cleaned, maintained, and serviced.

- *Standards and regulatory compliance.* It is essential to define applicable standards and regulations prior to detailed design up front, to permit efficiently addressing these paramount issues in the normal course of evolving design.

As mentioned, the marketing spec effort is led by marketing, but technology plays a substantial role too. Marketing and technology must work together. Marketing depicts a dream product in the language of the intended user (see Figure 9.2), and technology is there to keep the marketers technically honest. Not only must the marketing wish list conform to the R&D product description and physical law, it also must be practical within current engineering capabilities and remain within company reason.

That last comment has a subtle kicker. It is best if the administration division is also represented in the marketing spec process. Keeping the project within company reason often can be far more constraining than physical law. The solution is to include administration in all multidisciplinary gatherings where business perspective is a factor.

9.2 The Design Spec

In a sense, the design spec is the opposite of the marketing spec. Here, both marketing and technology are represented again, and to play it safe, administration should be present as well, to impose sound business perspective. However, technology now takes the lead, with marketing on guard to keep things honest. The reason is clear. The design spec translates the marketing spec into the language of engineering.

Translating critical information from techno-language to user language (the marketing spec) or from user-language to techno-language (the design spec) is a fascinating study in itself. Having taken the lead in complex R&D projects, we have managed these daunting processes firsthand. Conducted properly, by people from all disciplines who exhibit mutual respect, it is productive, rewarding, and fun. Still, misunderstandings can arise and persist, which virtually ensure eventual product problems, as demonstrated in the following categorical examples by company division:

- *Administration.* A broad weakness is poor appreciation of the tremendous work and creativity involved in both technology and marketing, exemplified by unreasonable time and content expectations and insufficient allowance for resource needs.

- *Technology.* Common problems often stem from poor knowledge of finance, funding, and business disciplines (administrative interactions) and the criticality of marketing research, marketing inputs to design, and technology's role in helping product marketing (marketing interactions).

- *Marketing.* Dedicated focus on the customer and product use can lead to unreasonable expectations regarding technology's required design and production time and costs and administration's monumental efforts in funding, providing resources, and creating the environment for company existence and functioning.

Technologists, marketers, and administrative types speak different languages, reminiscent of the "he's from Mars, she's from Venus" paradox in spades. Patience, understanding, and extreme attention to precise cross-discipline interpretation of all dialogue are the key to success.

Figure 9.2 is an example of a design spec derived from an accompanying marketing spec for the same product. First, note the item-by-item translation, every item in the marketing spec being specifically addressed (individually or collectively) in the design spec. That emphasizes the importance of ensuring that the marketing spec is complete, because it forms the topic superstructure for all other product-related specs. Also note that the design spec converts the user-framed nature of each marketing spec element into quantitative or at least technical boundary conditions appropriate for engineering.

The described R&D, marketing spec, and design spec process serves the grand purpose of control. Properly executed, the sequence ensures that issues of uncertainty, contention, or misunderstanding between marketing and technology are resolved as follows:

- *Uncertainty.* The root of uncertainty is marketing expectations that are unstated or ambiguous, a situation managed by making sure the marketing spec is complete and that all contained issues are clearly stated.

- *Contention.* The suggested design sequence is intended to highlight any points of disagreement, so the issues can be efficiently addressed and resolved early in the design sequence.

- *Misunderstanding.* This miscue category revolves around the distinct foundations of knowledge and experience between technology and marketing. Participants must be highly sensitive to the problem and exercise great care in interdisciplinary communications to ensure mutual understanding and interpretation.

The extra marketing spec effort as detailed design begins can save a fortune in later redesign and corrective action. Doing it right the first time usually means doing it faster and cheaper in the long run, and the hidden bonus is that everyone is still friends afterward.

It is important to have confidence that the marketing and design specs are complete. There is no way to be absolutely sure, but with the proper marketing expertise on board, all key user issues should be identified in one form or another. Extra assurance arises when you effectively meld skilled marketers and technologists together in the extended R&D, marketing spec, and design spec process. If the interactions are truly working, the technical people will force the user-framed thoughts of marketers into meaningful engineering restatement, and the marketers will draw technologist ideas in line with real user needs. Given sufficient dialogue that is functional, and assuming the proper expertise, the collective effort very likely will render effective solutions. Of course, different groups may reach unique conclusions. That is what competition is all about.

One technique often associated with the design spec step is the block diagram (Figure 9.3). This summary technique of system description breaks the entire product into a set of individual subsystems called blocks. All flow of data, control signals, forces, and other influences acting between the outside world and individual blocks and directly between the various blocks is diagrammatically detailed. The block diagram helps the viewer visualize overall design and component relationships, features that the design spec conveys poorly.

The block diagram is a potent tool for engineers, but it is especially powerful in tending to marketing and technology interactions. Being visual, the block diagram undoubtedly helps the marketers more than the technocrats. The technology people are familiar with picking out interrelationships between subelements of complex systems, because that is their training, but the marketers need all possible help in picturing what the design folks have in mind. The block diagram is simple in concept, serves to organize large amounts of information in concise form, and greatly assists multidisciplinary interchange.

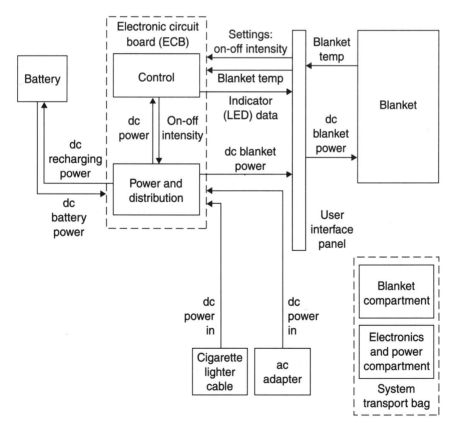

Figure 9.3 Block diagram of same product as in Figure 9.2.

9.3 Detailed Design

At long last we arrive at product design. There is an interesting relation here, unnoticed by even many technical people. For reference, compare the length of this section on detailed design to the length of Part III (Chapters 8–11). Formal detailed design is but a minuscule fraction of the total product development effort.

In any case, detailed design is where the specifics are imagined, formed, rehearsed, analyzed, tested, and hammered into the final product. In reality, the output of detailed design is not the product per se, but paper or computer files. Designs are documented descriptions, not products. The design is a complete set of written and graphic descriptions of every product aspect necessary to make and understand the product. Designs are abstract, while products are tangible, even a software product being physical and palpable in the form of its eventual sale.

The process of detailed design develops precise physical embodiments that fully materialize the specifications defined in the design spec and block diagram. These technical specifications completely convey the spirit and intent of the functional specifications defined in the marketing spec.

This book is not about detailed design, but we want to introduce a truly outstanding model system around which to organize a design operation. In fact, the model extends far beyond detailed design to outline a comprehensive quality, design, and manufacturing program.

The official program source is the Food and Drug Administration (FDA) of the Public Health Service, U.S. Department of Health and Human Services. The program is introduced in the *Medical Device Quality Systems Manual: A Small Entity Compliance Guide*, issued by FDA's Center for Devices and Radiological Health. The publication describes what is informally called the Current Good Manufacturing Practice regulations (CGMPs). Detailed sources and contact information are presented a little later.

It may seem odd that we have selected an FDA program as a model, but if your product has any medical markets, you will be pleased, since FDA clearance for marketing is required for any product that carries one or more medical claims. Still, the real reason is that the CGMPs possess a wealth of features prudent for any manufacturer. It is just that if your product is not medical (or military, aeronautical, or some other regulated category), you are not obligated to comply.

The CGMPs define requirements for comprehensive control of myriad company activities that clearly show what regulated industries demand. In the words of the FDA, the CGMPs

> govern the methods used in, and the facilities and controls used for, the design, manufacture, packaging, labeling, storage, installation, and servicing

of all finished devices intended for human use. The requirements . . . are intended to ensure that finished devices will be safe and effective and otherwise in compliance with the Federal Food, Drug, and Cosmetic Act.

Here is the value for you. A collective of top minds in the country have worked together more than 20 years to develop the CGMPs. The result boils down the entire design, quality, and production process into a standardized specification for overall performance (a performance quality spec). The CGMPs resemble a marketing spec, telling you what to accomplish, not how. As long as your system gets the job done with adequate assurance, the exact method is left to each producer.

If your product is not regulated, you can be selective, using the CGMPs as they stand, using just some of them, using some as is and adapting others, or not using any at all. The choice is yours. Even so, our advice is to employ the CGMPs, tapping the tremendous wisdom inherent in this monumental effort. We suggest you basically follow the CGMPs (described in Chapter 10), perhaps minimizing the meticulous details of records and other system elements in non-critical areas. Thus, even following our recommendation, there will be enormous variation in CGMP compliance. Some products with much critical content probably should use the CGMPs literally, whether the product is medical or not. Other products may contain little or no criticality. Level of adherence is a decision for company management.

In addition to the specific CGMPs, a wealth of useful information is available free from the FDA. Two direct sources are:

* FDA Division of Small Manufacturer's Assistance: (800) 638-2041;
* FDA World Wide Web home page: http://www.fdahomepage.html.

You can directly download the entire CGMP regulation and a host of other documents from the FDA Web site.

Throughout the book, our view of the product design cycle and its control adheres to the CGMP spirit, but we will not get deep into the detail. For one, that would take us far afield from the theme of business development, but book-relevant projects also vary far too widely for an in-depth tour. Our goal is to paint a broad picture of business development, much like the CGMPs do for product development.

The relationship of the CGMPs to an alternative and more publicized quality management system, ISO 9000, is critical. ISO 9000 is a set of international quality management standards, introduced in 1987 and since revised and widely recognized, coordinated by the International Organization for Standardization in Geneva, Switzerland. Both the CGMPs and ISO 9000 admirably accomplish essentially the same objective, providing a comprehensive standard for carrying out product development, design, production, and distribution that details performance requirements for virtually all major technical and some additional company activities. We have chosen to emphasize the CGMPs in this book because of their ease and economy of implementation, the material is free, and degree of compliance and adherence levels (assuming nonmedical products) is the company's choice. ISO 9000 compliance can be addressed similarly, but meaningful pursuit requires formal registration with an ISO-sanctioned registration body (e.g., UL), an expensive, perpetual association.

9.4 The Product Spec

The output of the last step, detailed design, is a set of engineering documents that fully describe product design and rationale. The documents do not represent the whole job by any means, as shown in Section 9.5. Before that, there is one last element of product documentation left to cover, what we term the product spec.

Figure 9.2 illustrates a typical product spec. The sample product spec represents the same device depicted in the marketing spec and the design spec of Figure 9.2 and also the block diagram of Figure 9.3. The product spec is a specification based on the detailed design for the finished product. It differs from the design spec, because much more detail is now known. The design spec sets boundary conditions and general technical direction for an evolved design concept, while the product spec defines details of all appearance, structural, and functional particulars of the actual physical product.

The main function of the product spec is recording the details of design, but overall it serves a number of prominent purposes:

- *Document the design.* The product spec summarizes the appearance, structure, and function of the overall product. Note that specs for individual block diagram components or other subsystem elements may be prudent as well.

- *Quality control standard.* The product spec or selected elements thereof can be used in manufacturing as a standard for assessing subassemblies or finished product.

- *Returned goods standard.* Goods returned by users or units in the distribution chain often must be checked against a uniform reference, to decide if rework is required.

- *Source of promotional data.* Marketing usually employs the product spec as a prime source of promotional material data. Said data may or may not be recast in the language of the user. This process basically reverses the translation of marketing spec to design spec.

- *Reference for planning.* The product spec is used by planning as a benchmark to assess product development cycle efficacy and as raw data to rate and improve the business planning process itself.

The product spec affects all three functional divisions of the company. Planning is an administrative function, and marketing governs promotional literature. The other three items relate to technology.

The following material outlines generic content for a product spec relating to an electromechanical hardware product:

- *Dimensions and weight.* Precise specification of the stated parameters is in order, for the entire system and for individual modules or distinct elements thereof.

- *Aesthetics.* Details of main shape, contour themes, intermodule aesthetic relationships, surface colors and textures, materials, and any other key aesthetic parameters should be detailed.

- *Primary functions.* The main product functions and respective ranges/ nuances should be quantitatively described here.

- *Secondary functions.* Details of any secondary functions or capabilities, useful to the user but not essential for product primary function, should be quantified.

- *Ease of use.* Internal or external design features or considerations meaningfully contributing to product ease of use should be explained in full quantitative detail.

- *Portability.* For products that are not usually left in one spot continuously, provide measurable technical details regarding portability-specific topics (e.g., transport methods and specifications, appropriate sources of power, special precautions) and any necessary issues relative to subsequent product use (e.g., setup procedures, recalibration).

- *Quality and reliability.* This category involves warranty information, acknowledgment of quality procedures used (if appropriate), and specification of mean time between failures (MTBF) and other measures of reliability.

- *Cost and price.* For company internal use only, it may be pertinent to list the estimated COGS at various levels of throughput and production experience (it is necessary to specify these two variables and any others that significantly affect cost efficiency).

- *Documentation and training.* Quantitatively described here are all elements of documentation, labeling, orientation, and training intended for salespeople, product users, or anyone else in the product distribution and use chain.

- *Cleanability and maintainability.* This category includes specifications for cleaning the product system and its elements and for all considerations related to repair and maintenance.

- *Standards and regulatory compliance.* Every identified compliance requirement should be defined and the chosen means of compliance quantitatively described.

Some of the technical product spec data discussed here are for internal company use only and would never be presented to product users.

9.5   Parallel Product-Related Activities

Detailed design usually is reserved to denote physical design of the product proper. Here we address important added components of overall product design.

9.5.1   User Manual

The most obvious ancillary item for many products is a system-use description. Low-end treatment might be no information at all or some form of hang tag or small label. High-end systems typically include a detailed manual or document set, including setup procedures, component descriptions, operating instructions, troubleshooting ideas, warranty data, and much more. Many modern products offer descriptive material of various levels in electronic manuals or as on-line help, alone or in addition to more traditional approaches.

It is easy to let the manual slip as the many hot spots of product development come and go. However, the manual is critical and must command due attention, since it usually serves as the entire interface between producer and user. Some Japanese and European consumer product manufacturers still treat the manual as an afterthought, so consumers have developed the habit of asking to see a manual before purchase. Manuals must be clear, written to the lowest level authentic user, organized and indexed efficiently, helpful in resolving problems, and functionally organized.

9.5.2   Training Manual

Complex products may require separate training manuals, and some may even need factory or field training sessions and perhaps user updates to address new applications or interpretations as the relevant field advances. Technology should lead the development of detailed content, but marketing should be significantly involved in defining the general approach and reviewing content detail. Marketing research should identify such needs with ample lead time for proper accommodation, and administration (planning) must arrange adequate focus and resources for this often slighted arena.

9.5.3 Maintenance Manual

Product maintenance and repair needs vary greatly, so marketing research should be carefully designed to include consideration of such requirements, if relevant. Supercomplex systems may even dictate on-site training or service representatives. Maintenance issues and the requisite handling of spare parts and materials are essentially in the domain of the technology division, but marketing people should be consulted to define exact user needs and desires. As always, the administration division (planning) must allow requisite focus and resources.

9.5.4 Technical and Regulatory Approvals

Here, demands range from nothing to overbearing. Both the technology and marketing divisions must diligently ensure that all required technical approvals are identified early. Examples include domestic (e.g., UL, FDA, military), state (e.g., water and air standards), city (e.g., water, air, and electrical criteria), industry standards (Health Industry Manufacturer's Association), and consumer groups (e.g., *Consumer Reports* magazine). No doubt some products will face other approvals. Know the requirements, all of them, and know exactly what is needed to comply.

9.5.5 Protective Actions

Check several legal maneuvers possibly relevant to protect the company's intellectual property (see Chapter 6). Protections include patents for products, copyrights for written or coded material, trademarks for item or process names, and service marks for names or designations of services. In addition, operational agreements control the relationships and obligations between separate companies and other legal entities. Finally, nondisclosure agreements restrain parties from using or disclosing information or knowledge provided in agreed confidence or control future competitive actions.

9.6 Prototyping, Testing, and Documentation

This section addresses three collateral activities associated with the design stage of business development.

9.6.1 Prototyping

Prototyping is often quite important during the design stage of business development. Many situations may dictate the need for some physical rendition of the product or a component thereof. Key prototype forms include but are not limited to the following.

- *Feasibility model.* Here, the idea is to produce some end result (say, turning on a lawn sprinkler system), but the activating mechanism used in the demonstration (human operation of a manual valve) differs from that destined for the finished product (timer activation of an electronically activated valve). The purpose is to show how one subsystem works before all subsystems are in place.

- *Functional model.* This model works the same way as the finished product. Interest may center on the whole product or on some isolated subsystem or mechanism. Appearance generally is not important. Common uses might include lifetime testing by engineering, user preference testing by marketers, or simple demonstrations by administrators.

- *Appearance model.* This is a full-size or scaled rendition of how the actual system (entire product or subelement) physically looks. This physical simulation also can include surface textures, knobs, displays, or other features, but the unit is not intended to function like the finished product at all. Classic examples are a software menu system that is not linked to any functions or a product shell to be used for promotional photographs.

- *Packaging model.* One purpose of a packaging model is to see how an array of known subcomponents can be assembled into a cabinet or other collective. Another reason is to study how an instrument would fit with other equipment and/or people, for considerations of size, access, interrelationship, or visibility. The model can range from a carefully machined version with additional uses (e.g., as an appearance model) to a stack of crude boxes (one to simulate each subcomponent) crafted from cardboard or Styrofoam sheets taped or glued together.

- *Preproduction prototype.* This rendition of the product is dimensionally and functionally correct but features components or assembly deviating from planned production units.

Note that all three company divisions may employ physical models at some point in business development, as exemplified by the following items:

- *Administration.* Feasibility models may be used to convince management of approach viability, and all model types may be used for demonstrations in funding efforts.

- *Technology.* The most prominent use of all five model types occurs during design development and detailing.

- *Marketing.* Functional and appearance models are frequently used in various promotional activities, as props for promotional material photographs, or as components of product exhibits at trade shows and other gatherings.

Planning must arrange adequate focus and resources for all requisite modeling.

9.6.2 Testing

Detailed design frequently employs testing. Some testing may have occurred in prior stages, but any previous efforts would have been directed to defining a design approach, not actual design. Several types of testing are common to detailed design:

- *Product operation.* An obvious evaluation is the performance of the finished product to verify and validate design or to aid the writing of various manuals and developing promotional material.

- *Compatibility.* This category concerns adherence to regulations, standards, and watchdog groups. UL standards and testing protocols (available for purchase) and national or industry standards (identified through technical or trade associations) are excellent sources of detailed technical information.

- *User preference.* R&D is the primary place to sort out user preferences, but often new questions arise during detailed design regarding how users approach, use, and envision products. This occurs especially with complex systems, such as complicated equipment or software, where user preference testing can involve an iterative test series to fine tune operator-sensitive design issues.

- *Subsystem operation.* Questions not serious enough to postpone design may still exist at design stage entry. Such testing may be conducted either in parallel with early design for incorporation later or after detailed design for inclusion only in later product revisions.

- *Lifetime.* Tests of product or subsystem lifetime are often desirable (e.g., industrial electronics) or obligatory (e.g., cardiac pacemaker). Lifetime tests statistically define existing performance characteristics and gauge improved lifetime design. True lifetimes of years or decades cannot be directly tested, of course, but techniques can be found in the literature to simulate extended time intervals using more practical spans (e.g., elevated temperature to mimic aging or intense ultraviolet light sources to approximate sun exposure).

- *Drop tests.* This item relates to handling and shipping, and drop tests are important for some products (e.g., cell phones) during use too. Relevant drop tests should be considered from the first inklings of design. Basically, the isolated or packaged product is dropped from specified heights at defined orientations and then tested for impaired performance.

- *Shipping.* This category usually includes drop tests but also involves evaluation of vibration, temperature, atmospheric pressure, and other variables peculiar to special shipping circumstances. Shipping tests can be simulated in the laboratory, where conditions can be precisely controlled. But a relatively cheap test is to actually ship one or more products or prototypes through relevant shipping channels, say, across the country and back, and quantify effects on product packaging, appearance, and function.

The listed tests are examples; actual tests and specific protocols must be tailored to individual product attributes. Also evaluate product packaging, labels and labeling, accessories, and any other associated elements, separately or in combination with product tests. Finally, the impact and compliance details of standards and regulations must be factored into relevant assessments.

We briefly mentioned that some tests must occur in detailed design by necessity, such as verifying feasibility of a particularly critical item in a group of interacting components. Other issues can perhaps be delayed for late detailed design or even later. One opportune interval for noncritical testing is the preparation for launch stage, the topic of Chapter 10.

9.6.3 Documentation

The output of detailed design is product-specific engineering documentation, not products. Here the view of documentation is broadened beyond purely product-specific descriptions, to formally list all prominent documentation categories that arise from the FDA CGMP system.

- *SOPs.* A companywide set of procedures that define main activities conducted within all three company divisions.

- *Files.* The CGMPs mandate certain files, and it is good practice to simply embed the required files in a companywide system of defined files covering all business, technical, and marketing activities not of a personal nature (consult legal counsel to define exactly how to define, categorize, and protect individual classes of file).

- *Drawings.* This family of elements comprises graphical product descriptions and representations ranging from the most minute detail to the entire assembled product.

- *Instructions.* This is a category of procedural specifications applicable to components, products, or their performance. Illustrations include manufacturing, quality control, and incoming inspection instructions.

- *Procedures.* It may be prudent to define a class of written procedural controls that lie below the SOP tier. Such procedures might address the operation of a particular piece of equipment, access control to a given room, and so forth.

- *Forms.* Forms serve to impose a set order and organization on a particular collection of related facts.

- *Specifications.* We have discussed specs associated with product design. But specs are ubiquitous, including those for processes, quality control measures, equipment, software, safety, health, and so forth.

- *Checklists.* This category includes defined rosters of technological, marketing, and administrative elements that serve as a reference of completeness.

- *Position descriptions.* At some stage, it likely will be prudent or necessary to formally define jobs or positions. Some key roles may dictate a more complex employment agreement. Samples can be found in the business literature.

- *Records and logs.* The CGMPs identify many instances where formal data needs to be collected and retained, to document the occurrence or nature of processes, actions, or results. Key players should keep a telephone log to register the time, participants, topics, and originator of all incoming and outgoing calls. As with the previous discussion of files, consult legal counsel to design a system of records and logs suitable for business purposes but minimally damaging in case of legal actions against the company.

The preceding list is not exhaustive, but notice that the items cover all three company divisions. Some said the computer age would eliminate paper, but it sure has not yet.

These details will be overkill for some projects, while other operations will need more elements than those mentioned for proper control. This book is a general introduction to business development presented in a broad nonspecific vein, so you must adapt book content to something useful but not burdensome in your realm. Concerning documentation, the key is to facilitate proper and efficient interchange and also C^3.

9.7 Design Reviews

It is crucial to review an evolving design at many junctures. Consistent with the CGMPs, this section discusses three classes of design appraisal.

9.7.1 Staged Design Reviews

We recommend establishing a formal procedure to assess the intent and content of a design at various product development milestones. Milestones should be defined during the design spec and block diagram effort, and all such activities should be conducted under a designated design leadperson, most likely someone from the company's technology division.

A typical design review meeting should involve the design leadperson (at the helm) and other key design people. Marketing must also be represented, commonly by the product leadperson from the marketing division, but administration is present only if the agenda dictates. Outside consultants may be retained to aid in certain design reviews, if deemed worthwhile. In any case, the design leadperson should send all divisions and review participants a written postreview summary.

Each design review must assess whether the design is on track. One way is to compare progress with the design schedule maintained by administration's planning function. Another benefit of design reviews is to get everyone synchronized on all major issues, since detailed work often leaves people isolated and out of touch with parallel activities. Also, design reviews help to document the design process, important to control the emerging product but also useful later as a model of how (or how not) to conduct future design. Finally, design reviews aid project advancement by instilling intermediate deadlines.

Many groups do not use design reviews at all. One common excuse is that the group is small and people interact daily anyway, while others say that design reviews are just one more incessant interruption to the true design effort. Both retorts are foolish. New concepts, unrecognized problems, conflicting specs, and an endless list of other issues arise during design reviews and nowhere else. Design reviews really make a difference, some plausible reasons including better and wider communications, a more formal atmosphere, and the pressure of more current rather than fewer remote deadlines.

In any case, design reviews are quite effective technically, and they undoubtedly save resources by coalescing thoughts and actions and keeping them on target. Even two- or three-person garage operations should carry out design reviews, the real problem being related not to small groups but large ones. Design reviews are not a social hour, and people who are not directly relevant should not be present. Design reviews with too many people are not only inefficient but conducive to serious delays of the meeting itself and of the overall product development schedule due to bombardment by needless concerns.

9.7.2 Design Verification

Formal staged design reviews have been depicted for predefined milestones during detailed design. When detailed design is complete, we then suggest two additional formal reviews. What we term design verification and validation really have more restricted purposes and formats than traditional staged design reviews during detailed design. We discuss design verification first and then design validation.

Design verification involves comparing the completed design (the design output) to the design spec (the design input). In other words, the charge is to see if the actual design meets the design input requirements.

Notice that design verification is limited. Issues of how to do this or that are not germane. The task is to strictly compare the design spec to the design, so the focus is very tight, helping to ensure that the evaluation gets directly to the heart of any outstanding or arising issues.

If the design verification meeting turns up problems, then the design is in essence rejected. Subsequently, the design group must address the problem and then call for another design verification session when fixes are complete. The discipline must be that the product design does not advance until the fix is in, so to speak.

9.7.3 Design Validation

Design validation is one last cross-check on design soundness and competence. While design verification matches the design spec (intention) with actual design (documentation), design validation compares real product performance to defined user needs and intended uses (the marketing spec).

So design validation must wait awhile. All we have right now, at the end of the design stage, is a paper (or computer file) design. We might also have prototypes of some kind, but not production units. We mention design validation at this point in the discussion for logical continuity with the other forms of design review.

9.8   Administration and the Design Stage

The burden on administration is always heavy, because planning, the business plan, and funding are the linchpins of everything to be and are the most demanding of the seven success factors. Still, as we have seen over and over, that is just one point of view. In the limit, there is no basis to credit any one company division, administration, technology or marketing, over the other two. The company is a triumvirate, the only form of sustainable existence.

Like every other business development stage, the design stage harbors special challenges for administration. The items listed here underscore administration's principal concerns during the design stage. The headings correlate with those in Chapter 8 describing administration's functions in the R&D stage.

- *Facilities.* The design stage sees mounting interaction among the three company divisions. Substantial dedicated company space may be needed, or, alternatively, much of design and perhaps other functions may be contracted, retaining a home-based team structure. In any case, the facilities demands can become pronounced, and it is administration's job to project such needs and make them happen. Creative facilities arrangements can often effect major cost savings, which investors relish.

- *Human resources.* There is work to be done, requiring either contracted or in-house people. Often, in-house people help hold down costs, especially if working in part or totally for equity. Regardless, administration must secure the needed resources, which means having sufficient people within a motivational, professional, and legal environment sufficient to accomplish current work and also accommodate the future. Money spent here is the hardest to justify to investors.

- *Nonhuman resources.* Common necessities can involve software, equipment, tools, parts, materials, special processes, outside services, and others. As always, the issue's magnitude is extremely sensitive to product nature. Administration again must weigh contract versus in-house action, in the context of past, present, and future needs and resources. Investors expect frugality in this category but understand the need.

- *Planning.* The advancing company is becoming bigger and more dynamic. Outside interactions are growing, too, and the demands for current and future funding probably are accelerating. Still, the R&D

stage should have created material for demonstrations, which now should be groomed and leveraged to the hilt. From here on, every new achievement should be exploited to build business plan strength and bolster funding efforts. You have entered the do-or-die era.

- *Funding.* It is fair to say that the R&D stage usually sets the stage for major funding. In the same light, the design stage is where large funding needs typically arise. Usually beginning with R&D, planning not only must ensure completion of the tasks of each stage but also create the base to fund the next stage. This is not magic but requires carefully orchestrated product and business development results and a credentialed, talented, and connected team to find and secure successive funding.

The challenges facing administration are monumental but manageable, and again we see the supreme criticality of building a powerful management team.

9.9 Marketing and the Design Stage

We close this chapter with the primary issues facing the company's marketing division. During the concept and R&D stages, marketing was called upon largely to support activities of administration and technology. Marketing helped administration through contributions to planning and to funding efforts and assisted the technology division in product conceptualization and R&D. These efforts were critical to company progress, but marketing was backstage, providing essential but largely invisible support.

However, come the design stage, the background role of marketing changes forever. Marketing still is not visible to the outside world, but it leads in-house development of the marketing spec, and marketing's role in mainstream design reviews and in design verification is apparent within the company. These and other features of marketing's design stage prominence will be revealed though the items in the following list.

The functions of marketing are evolving with time, so the headings here differ somewhat from those of the R&D marketing discussion in Chapter 8.

- *Product features.* The marketing division plays critical roles during design. Marketing directs the first step of design, the marketing spec, and is essential in the marketing spec translation that creates the design spec. Also, a concerted marketing role is imperative for all conventional and special (verification and validation) design reviews. In addition, marketing must always watch over the unfolding design and continue to examine relevant design issues independently of the technology division.

Classic instances where marketing can and should get involved include the design of user interfaces; product appearance, size, configuration issues, and packaging design; access to user interfaces and internal subsystems; and manual content and layout.

- *Marketing and sales plans.* The marketing division must develop the detailed plans for both marketing (customer characterizations, program design, product imaging, promotion, pricing, forecasting, distribution, customer service) and sales functions. Plans must be cost efficient, but marketing and sales plans that are too lean will not work, unless sales forecasts are greatly lowered accordingly. That is a bad idea too, because the results likely will not provide enough revenue for all company functions. So the target is a lean, efficient operation that is otherwise fully capable of meeting reasonable planning goals.

- *Contributions to planning and funding.* Enhanced marketing inputs are obligatory for effective planning and successful funding. Other team members cannot take the place of marketing expertise at any time, and marketing inputs are becoming all the more important as firm design element commitments accumulate. In addition, the pressure of funding needs likely has risen to commanding priority, and potential investors always want direct access to the marketers, to assess whether the marketing and sales plans are plausible, sound, and complete.

- *Marketing research.* Marketing research ideally should never stop. Here the important actions are to reaffirm selected markets, strengthen customer profiles, and build marketing data on imaging, promotion, pricing, forecasting, distribution, customer service, and sales as a basis for improved future planning. Also, if suitable prototypes are available, they should be used to enhance marketing research and to begin developing specific marketing channels and individual customers.

- *Product introduction.* During the design stage, it is prudent to sketch a preliminary plan relative to product introduction. The official product introduction stage is some time away, the preparation for launch stage coming next. However, product introduction is the pinnacle of milestones, its importance far exceeding stage duration, so an early start of meticulous preparation is a must. It is also advisable to conduct a series of marketing reviews, analogous to the design reviews of the technology division, with all three company divisions attending, to assess marketing plans as they become formulated.

Beginning with the design stage, marketing comes into prominence, a process even more evident in the next stage, preparation for introduction.

10

Launch Preparation Stage

10.1 Formal Launch Plan and Funding

Having previously detailed the conceptualization, R&D, and design stages of business development, the product has been designed by the technology division, and the administrative and marketing divisions have stayed abreast with their contributions. It is time to prepare in earnest for impending product introduction, a readying interval we have termed the launch preparation stage (or, more simply, the preparation stage).

We chose to impose another business development step, preparation, between the design and introduction stages for several reasons. After design, start-ups often rush the product to market, so placing a formal stage between the two flagged milestones may encourage valuable pause. Also, there are key activities that cannot start until design is over but that take time to complete. In addition, the interval allows the technology division to do some product testing and for the marketing division to gain a headstart on product awareness and name recognition. Finally, administration can use some extra time to line up the significant funding often required for the introduction stage. This chapter is about all of those points and much more.

This section takes up the merciless yet ennobling chores of planning and funding, requisite essentials to the entire business development process but especially critical during the preparation stage. The reason is momentum, because with the approach of product launch, any problems can become serious threats, and the more momentum builds, the more dire any disturbance becomes.

The basic rub is costs. There are a lot of them, and they just keep coming faster and faster. The onslaught really never ends, but if the product launch is successful, revenues begin to flow too. Finally, if planning is accurate or if the mad scramble to keep the project afloat anyway happens to work, revenues must be programmed to eventually exceed costs.

Still, the company must survive in the meantime. To effect that end, administration must use planning and the business plan as major tools in the act of searching out and securing funding. The dynamics of planning and funding are urgent during preparation and during introduction as well. Thus, it is instructive to revisit the topics of planning, the business plan, and funding, subjects central to every business development stage but carrying special considerations during the preparation and introduction stages.

10.1.1 Planning

Planning was the central subject of Chapter 5, which presented an in-depth look at eight purposes of planning. It is time to reassess each cited reason, along with certain subtleties related to the preparation stage.

- *Build the management team.* By the end of the preparation stage, the management team typically should be intact. If not complete, it is critical to fill out the team and to ensure that the group remains whole, to both be and look the part of a winner as the company enters the introduction stage.

- *Set goals and priorities.* Goals and priorities are used to develop the schedule inherent in the financial model. The model is critical at this point, since company dynamics are high and costs are substantial, so accurate modeling is at a premium. Planning also must constantly monitor predicted versus actual progress, because rapid corrective action to any problems is paramount.

- *Compare past and future.* The past is consulted all through business development, to aid planning and execution, and there are no special pressures or considerations of import during preparation.

- *Forecast and meet human needs.* The mounting rapidity of change and workload during preparation can call for particular emphasis on staffing. The key is not to relax hiring criteria to fill the need for people; ineffective hiring leads to more problems and inefficiencies later.

- *Forecast and meet resource needs.* The demands for nonpersonnel resources can be acute but are known far in advance. Arrangements in at least preliminary form should be made well ahead of actual need.

- *Model future operations.* This model was mentioned in the item "Set goals and priorities." The emphasis here is on carrying the model forward in time to predict future needs long before the requirements appear.

Frequent model updates are also in order, given the highly dynamic environment.

- *Disseminate information.* As dynamics pick up, C^3 becomes all the more vital. One critical element to instill and sustain C^3 is frequent and effective communications, both within the management team and between management and everyone else.

- *Measure performance.* Planning must always track, analyze, and improve performance and productivity. The preparation stage is not special, but it is not a time for relaxation either. It is always prudent to augment productivity through supplying better equipment, computers, communications, and other relevant factors.

Each of the prime purposes of planning is still quite important during the preparation stage, and several of the items are especially prominent.

10.1.2   The Business Plan

The attention now turns from preparation stage planning to considerations of the business plan derived therefrom.

Chapter 6 provided a detailed study of the business plan, focusing on content, structure, and physical makeup. A number of specific purposes of the business plan were developed, and we now reexamine those needs as a refresher and to discuss special aspects of the business plan during the preparation stage.

- *Funding.* When needed, funding is always of the utmost importance. Even so, the needs are likely to be far greater during preparation and on into introduction, and the intervals between funding needs may be unusually short. Finally, funding can be more urgent in the sense that project delays or lapses may be more damaging.

- *Frame of reference.* The business plan has multiple purposes at all project stages, but the preparation stage is so dynamic that tools to aid consistency across distinct company functions acquire added import. Along with C^3, the business plan is the most obvious entity available to coherently define and channel actions.

- *Out-of-house communications.* The business plan is almost always the best tool to convey status and intent to vendors, key customers, and business professionals outside the company. The fast-paced world of preparation accentuates the need for a superior plan that is also frequently updated.

- *In-house education.* The accelerated dynamics of preparation may raise demands for educating some or all supervisory and general employees. The business plan is an excellent instrument of education with regard to present and future aspects of the company and its subelements.

- *SOPs.* Planning sees the business plan as a potent standard for measuring company, division, and function performance at every stage of business development, not the least of which is the highly volatile preparation stage.

- *Means to track planning.* The business plan not only permits assessing performance, but with time, the company has more history from which to learn. One ongoing focus of planning should be to improve itself (its own methods and procedures). Do not let the fast-flowing dynamics of preparation dissuade constant striving for improvement in planning and all other key activities.

- *Departmental plans.* The sometimes blistering rate of events and demands places high priority on forceful and careful planning by individual departments. This imperative arises due to each department's charge of correctly implementing, on time, their respective components of the overall business plan.

Remember that the business plan for the whole company forms the master gear against which all departmental plans must mesh precisely.

The high dynamics of the preparation stage should not be allowed to excuse or let slide ongoing essentials. Rather, employ the faster pace as the reason to fix and tune planning and operations.

10.1.3   Funding

Chapter 7 described a stepwise process to seek funding, an approach applicable to any business development stage. Nevertheless, different stages also dictate unique considerations. This section recaps each step of that prescribed funding process, for review and to highlight preparation stage peculiarities.

- *Lay out the plan.* At every funding stage, it is wise to assess the situation independently and to design a new funding plan accordingly. Things change, especially for funding, and the preparation stage is probably not the first funding you have sought. Thus, you have useful new leverage in the form of company progress and respective artifacts and plans since prior funding efforts.

- *Verify all is ready.* You must ensure that the company and the management team are completely ready with the entire funding package and predesigned series of presentation steps. This must be done prior to even preliminary talk of new funding with anyone. On the basis of interactions to date, decide whether to give existing stockholders an early, special shot at the new offering.

- *Test the plan.* This is a huge funding round. You have been successful at funding before, the project is well down the road, and introduction is near. You are probably in a strong position to have the business plan critiqued by various professional and formal groups, such as those discussed in Chapter 7.

- *Build a list.* One early task is to create a list of contacts. Preparation stage funding is critical, so use a group of noncandidates to test your presentation. Include prior investors on the list, if they are considered candidates again, and realize they also may be able to suggest additional contacts that they were hesitant to mention earlier due to less project advancement and success. Additionally, use all the other ideas mentioned in Chapter 7 to flush out candidates.

- *Set up tracking.* By now, the company should have an established routine for tracking investor candidate contacts and a record of how each candidate has progressed with time. Adjust the method to incorporate improvements learned from experience and use it again.

- *Initiate contacts.* Once all is ready, begin the contacts, taking care not to attack the list either all at once or one contact at a time. All at once is too fast and does not allow for experience-based tuning. One at a time takes forever, because each candidate may play out slowly. Take the candidates in small groups, employing the method depicted in Chapter 7. Also, pitch the project from a position of strength and confidence (but not arrogance), since you represent a project that has a strong and skillful team, a fully designed product with identified markets, and a company that is nearing product introduction.

- *Corrective action.* As the contact list is being addressed, carefully note how the project is being received. One instance of negative vibes is not usually of deep concern, but patterns of reaction, good or bad, are critical. Do not be passive; ask people for comments about the business plan or other presentation factors. Do everything possible to identify what is good and how to strengthen what is not.

- *Contact followup.* The discussion of contact followup in Chapter 7 is completely representative of needs in the preparation stage. Followup is beneficial in all sorts of human interactions, because people acquire identity and develop stronger ties in proportion to exposure and rapport. Certainly play to the interested, but cousin the disinterested as well. Elements of the latter group may be valuable in the future, if for no other reason than name and product recognition.

- *Persist.* You must incessantly maintain the effort. If success does not come early, keep the pressure on and your spirits high. Nothing has changed, and the project is still as strong as before—you just have not found the right people yet. Investors are as strange as the rest of humanity. You have to catch them at the right time relative to general mood, family and business comfort, nature of their holdings, economic cycle, tax obligations, other purchases, the sale of separate assets, and so forth. You must find not only the right person but the right person at the right time.

Funding is difficult. Take it seriously but also realize that you are not begging. If you have followed our recipe for business development, you have a valuable commodity. You are offering a well-designed project that has a track record of success and a solid plan for the future. Be proud—you have earned it.

10.2 Administrative Preparation

Administration can use the preparation stage to great advantage. For example, numerous important company internals must be developed at some point. Of course, it is best to design and install these systems as early as possible. We describe them here because the preparation stage provides a logical opportunity.

Administration is busy during preparation but less visible than the other two divisions. Activity and emotions until now have focused on creating a product and finding markets. However, putting the company in order cannot wait any longer. As an illustration, organization renovation as outlined in upcoming material augments company depth, which, in turn, can aid greatly the tough funding efforts of this stage, preparation, and the next stage, introduction.

So this section and the next two will show how the company should be remodeled. After all, the project has come a long way, and complexity has grown immensely.

Chapter 9 introduced the CGMPs, a mandated quality system (largely analogous to ISO 9000) imposed by the FDA on medical device manufacturers. The CGMPs convey profound wisdom and utility for any product-oriented company. The ideas presented here are adopted from the CGMPs. We have carefully carried over CGMP terminology to enhance correspondence, even though our treatment is a cursory CGMP abstraction. We highlight some of the best of the best but at the expense of much useful detail, so reference the CGMPs as well in designing your system.

The first issue affects all three divisions of the company, administration, technology, and marketing. It is time to carefully restructure the company in light of growth and foreseeable need. Two points are critical. First, do not reinvent the wheel, since the literature is full of excellent examples relating to virtually any major issue. Second, we recommend the CGMPs as your system's structural base, tapping the tremendous wisdom contained therein, in a way consistent with this book.

The CGMPs may seem complex at first, but they really just outline central elements required for a functioning, coordinated operation. Figure 10.1 depicts the general structure of a typical product-oriented growth company, listing all normal functions. In turn, the following material is more selective, emphasizing elements singled out from normal functions that particularly resonate with the CGMPs. Folding the CGMPs into the ongoing company function and culture of Figure 10.1 is the purpose of suggested company restructuring. The key is the word outline. The CGMPs do not specify how to implement the defined elements at all, so you are free to define detailed relationships and functions. Your system may be simple, or special needs may dictate something much more complicated, a choice that is entirely up to you.

We must make a point about all the functions discussed in this section and in Sections 10.3 and 10.4. We will be addressing specific functions, such as the office of the president, planning, quality control leadperson, and so forth. This is a general-purpose introductory treatment of business development, addressing all sizes and complexities of project. Thus, when we refer to the president, the range of reference is enormous. In your case, activities of the president may be a full-time effort. Alternatively, the entire project may involve just the original founder and one to several other key part-time players working off-hours. We are depicting functions, not people. References to the president mean the respective activities and functions of administrative leadership, not a person.

Now let's step down from the level of the overall company to look at administration division activities, splitting it into its four constituent parts.

10.2.1 General Management

Requirements for general management have existed all along, since the company was formed and has kept functioning to date under someone's lead. Still, if not recognized already, it is time to formally define a general management position and to designate a leadperson for this function, typically the corporation president.

The president is in charge of all company functions, reporting to the board of directors, a formal group elected by the stockholders. In turn, the

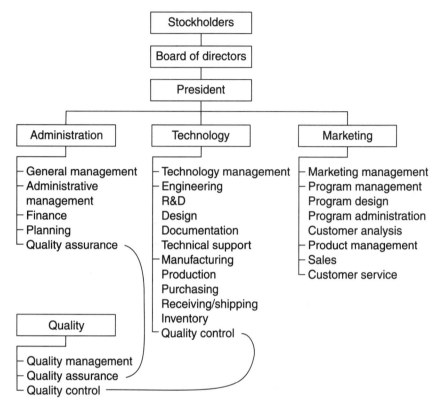

Figure 10.1 Diagram of generic product-oriented company structure. Stockholders own the company and elect the board of directors, who in turn elect the president. Functions are organized according to the three divisions emphasized throughout this book. A fourth division, quality, is also depicted, assuming control of functions previously in the administrative and technology divisions. (Formation of a separate quality division is recommended in Chapter 12.)

board appoints the president (and perhaps other officers) and sets general company direction (strategy) within the governing rules of the company, the bylaws. The president is charged with running the company under guidelines defined by the board, personnel management, legal matters including agreements with various outside entities, and all three of the other titled functions (finance, planning, quality assurance).

10.2.2 Finance

Finance is meant to include maintaining the monetary books, supporting materials, and records for all company activities. Usually a company hires outside accounting expertise for consultation, preparation of major filings, and any required audits associated with funding efforts. Finance does not involve modeling, which is an activity of planning.

Most likely, you will need outside accounting expertise, and it may be an advantage to retain one of the big national accounting firms for that purpose. They certainly can provide the accounting help and any required advice, although probably at a high price. The advantage is that such firms may be able to help with funding in two ways. First, the company looks more substantial and stable if coupled to top-drawer professionals, including accounting, legal, and perhaps insurance firms. Second, such firms may have clients that are potential investors (although you must also ascertain whether they are willing to inform any clients about your project). Each project must weigh the potential benefits against costs.

10.2.3 Planning

Here we are lumping planning, business plan maintenance, and funding into one formal administrative subgroup called planning. Those functions are such essential subjects that all of Chapters 5, 6, and 7 were devoted to them. Suffice it to say here that planning, the business plan, and funding are never more critical than during the high dynamics of the preparation and introduction stages. One major reason for defining a separate preparation stage was to allow time and to flag the importance of accentuated planning, business plan, and funding preparation and action.

10.2.4 Quality Assurance

The last category probably will not be found in many other books. As far as we know, the definition of a distinct QA function is our own concoction.

Intrinsically building quality into a product rather than trying to inspect away problems in later production saves money. We are not aware of any published exceptions, but the same is true of any activity. Do it right the first time and save all the trouble of rework, an assertion difficult to refute. Let's carry the argument one step further, by suggesting you design quality into your whole business, not just its products.

We propose that you install what we will term a QA function, an act correlating well with the CGMPs. QA reports to the president and no one else. QA directs all key functions of the company to create and use a set of SOPs. The SOPs represent a written description of every main activity within every nontrivial position in the company. Many successful product-oriented companies already have such a compendium.

The SOPs are very useful for training and as a guide to action. Additionally, the SOPs give general management, through QA, a powerful tool to periodically assess and improve performance. QA is charged with assessing all company activities, by conducting audits comparing actual performance to the written procedures. Under administrative control and approval, QA then designs corrective actions and ensures that they are put into actual practice.

10.3 Technical Preparation

This section examines how the CGMP theme applies to the company's technology division. The CGMP requirements are intended to ensure that all identified functions and processes are properly managed, maintained, conducted, documented, and reviewed and also that functions and procedures are controlled and updated as needed to ensure effective results. The headings in the following list relate to comparable headings in the CGMPs.

- *Design controls.* Here the CGMPs set forth design requirements discussed in Chapter 9. Topics include design inputs, outputs, and reviews; design verification and validation; and certain design records. In short, design intent must meet appropriate standards and be complete, and ultimate design must correctly conform to design intent.

- *Document controls.* Here the CGMPs treat how design, manufacturing, and quality documents are created and how document change must be handled. Original documents must be approved by proper authority and distributed so that correct versions are in use and obsolete documents are not. Document changes require the same approvals and distribution controls.

- *Purchasing controls.* The intent is to ensure that vendors are properly approved, received components conform to requirements and are properly controlled, and detailed purchasing records are maintained. Purchasing data that specify product specs must be under formal document control.

- *Identification and traceability.* The manufacturer must be able to identify product at any stage of receipt, production, and distribution, to prevent mixups and to aid recall of problematic materials, parts, and finished product. Usually, lot number control is sufficient, but serial number control may be required in critical cases.

- *Production and process controls.* The manufacturer must develop, conduct, control, and monitor production processes, to ensure that the product conforms to spec. Also, nonconforming parts, materials, or product must be handled in ways that ensure proper segregation and control. Facilities, workers, contamination, and the work environment are also to be tightly managed. Additional requirements include control of production equipment and its maintenance and calibration, and of process controls and process validation.

- *Acceptance activities.* This includes inspections, tests, and other verification activities concerned with received goods and with in-process or finished product. Requirements center on the involved activities, tagging of relevant items, and records management.

- *Ancillary activities.* The CGMPs include additional sections for the following: nonconforming product; corrective and preventive action; labeling and packaging control; and handling, storage, and installation. The call is for specific written procedures that are properly designed, managed, and reviewed and for keeping proper records.

- *Records.* This defines the required nature, content, method, and location of storage; access; and control for key documentation. Several classes of document are required: product master record (design specifics), product history record (historical design progression), quality system record (QA system design), and complaint files.

That ends our whirlwind tour of the CGMPs. Some less prominent elements were left out, and CGMP details were barely touched, but the value of the CGMPs should be clear.

The CGMPs lay out general guidelines that should form the basis of any sound company, and investors should be favorably impressed by an operation designed to and controlled by such standards. The preparation stage is often an excellent time to design and implement SOPs and other elements of a company-wide control structure.

Another activity that frequently fits the preparation stage mold is formal and informal field trials.

- *Formal field trials.* Tests conducted under a structured protocol to define one or more specific design, manufacturing, or quality issues.

- *Informal field trials.* Informal tests to answer outstanding technical questions.

Field trials are strictly limited to technology division activities and are often integral to the total design process, necessary to answer questions unknown or not addressable during the R&D or design stages. An example is design validation of a medical product, involving a formal demonstration that actual product performance meets defined user needs and intended uses (the marketing spec). The CGMPs mandate that such tests be conducted on a first group of product from the authentic assembly line, prototypes not being acceptable.

10.4 Marketing/Sales Preparation

We now shift attention to marketing during the preparation stage. The extra stage between design and introduction is useful for the following marketing activities:

- *Product awareness.* Exhibits or private showings at technical meetings and at trade and market shows may allow important early exposure. Press releases to newspapers and magazines also should be considered. The goal is to build advance company and product name recognition.

- *Early promotion.* Beyond private showings, some product and market combinations lend themselves to early promotion, in the same or different media than planned after introduction.

- *Trade publications.* Most industries have trade newsprint publications that feature product information. Many provide free space for new products, and cost for small ads is reasonable.

- *Product reviews.* Some publications accept sample product for review. For information on this and other applicable publications, contact relevant trade or industry associations.

- *Training.* Complex products require experience or training for competent use. The preparation stage can be used to build familiarity and interest with key customers or reviewers.

- *Demonstration units.* Related to training is getting demonstration units to prime salespeople for orientation and experience.

- *User trials.* Marketing often can selectively place prototypes or production product before official introduction. This may garner testimonials, scientific publications, reviews, word-of-mouth exposure, and early evaluations for later purchases.

- *Marketing and sales material design.* The preparation stage is a golden opportunity to carefully complete the design and production of all requisite materials.

- *Marketing research.* Ideally, marketing research should never stop. Preparation stage efforts can be used to further check market assumptions and to refine customer descriptions and knowledge of marketplace buying systems. This can be critical, because many buying systems contain multiple players (users, bosses, administrators, purchasing departments, and so on).

Note that some of those activities listed might be unwise in certain markets to avoid alerting the competition early. For example, any type of premature private or public exposure of unique product features or promotional descriptions thereof can provide competitors with extra response time.

10.5 Simulated Program Introduction

It may be useful to simulate the introduction of your product, to evaluate alternative approaches and refine process details. A process can be imitated in one of two ways, employing authentic objects in a real-world environment or using some form of computer model.

Besides simulations, useful histories of product introductions may exist in the literature, the key being to find relevant examples. The point is to use every possible avenue to better prepare for any eventuality.

10.5.1 Real-World Simulation

You cannot realistically imitate the entire activity complement of bringing a product to market, because the most important element, real potential customers, are absent. Even so, there are aspects of the planned introduction that can be subjected to dry runs.

Depending on your product and market, the following functions may be usefully simulated to varying degrees.

- *Customer service.* This refers to simulating incoming telephone, fax, e-mail, and Internet contacts, including general inquiries, orders, requests for technical support, comments, and complaints. People who will tend the calls can develop scripts for various call types and then practice calls with each other. Some calls will require referrals to other company functions (e.g., technical service inquiries).

- *Orders.* Orders usually are recorded first on a paper or computer form and then pass through some defined process, ending in an archive file. The pathway is undoubtedly branched, to contend with order features or choices. Make up sample orders that cover every variance and run them through the planned system.

- *Processing.* The first two functions relate to the marketing division, but orders pass to the technology division for total fulfillment. Run the same sample orders used previously through every technology check point, to ensure that all system branches are accosted.

- *Shipping.* The test should pass through final product inventory (if relevant) and also packaging and shipping. Be sure the tests assess proper labeling and addressing, protections against order or address mixups, and the efficacy of tracking systems and record keeping processes.

- *Installation.* Installation will not apply to many products, but it is good to imitate unsupervised customer setup even for simpler systems. Demanding installations should be simulated, and any required training of installers should be appraised for content and efficacy.

- *Technical support.* Many products will require formal technical support, occasionally by customer service personnel but more commonly by specific technology division personnel. In the latter case, customer service must route the call to either a formal technical support function or a designated person on call. Both customer service and technical support should be constantly attended to maintain a professional appearance.

- *In-person promotion.* Some products will be amenable to shows, exhibits, demonstrations, and other forms of live presentation, and relevant public activities should be rehearsed to ensure smooth execution.

Even that entire collection of trial runs cannot prevent all problems, but it almost certainly will improve results.

10.5.2   Computer Simulation

Computer simulations can provide useful insight, but they cannot accurately imitate the entire product introduction process.

The value of computer simulations is mostly in exploring the financial dynamics of the product introduction process, using financial modeling described in Chapters 5 and 6. In fact, using several related models is best to depict various degrees of product acceptance and sale. Using proactive alternative response plans is a superior approach to merely reacting to happenstance. Alternative approaches to in-house spreadsheet modeling include hiring a consultant or locating special software to accomplish more elaborate simulation of preparation stage activities. Special-purpose modeling is beyond the scope of treatment here, but valuable direction can be pursued through high-caliber university business schools or libraries.

10.6 Commitment to Introduction

Preparation stage activities have no obvious end, but issues not previously addressed may significantly influence preparation stage duration and timing.

Actually, forces impel the start of introduction more than an end to the preparation stage. For instance, some products are strongly dependent on the season or specific holidays. Others may be timed according to relevant technical meetings, exhibits, markets, or product shows; for embryonic companies, those factors can represent critical deadlines lacking any leeway. In addition, products exist with pronounced dependence on the economic cycle (high-ticket luxury items), political tides (election seasons), and social sensitivities (environmental products). Also, timing may be critical relative to the actions of competitors (new product introductions, updated models), an issue or problem specific to the company (a window of opportunity), or its key personnel (family issues).

Instead of outside forces, company internals can also compel a logical move to introduction. Administrative reasons might relate to pressure from or promises to existing stockholders, meeting predefined schedules (which nearly always should be considered an imperative), or hitting a perceived window of opportunity. In turn, pressures with a technology bent might be related to agency or regulatory approvals or to upcoming changes in regulatory or standards environments. Marketing could be compelled by large new customers coming on line or other opportunities for high visibility or sales leverage.

One last force often overwhelms—the incessant need for money. Administration's ongoing funding search and the intricately tied activities of planning and the business plan commonly generate considerable pressure to complete preparation and move on to introduction, since preparation stage money usually comes with assurances of timely product launch.

So, one way or another, the company is typically compelled or propelled into the introduction stage, a monumental happening to which we now direct our attention.

11

Product Introduction Stage

11.1 Actual Program Introduction

Prior discussion has proceeded through the conceptualization, R&D, design, and preparation stages of business development. We have arrived at the pinnacle of focus for most entrepreneurs, the product introduction stage (introduction stage, for short).

Let's first define the introduction stage by identifying a set of activities associated with introduction. Some singular event, like the day of product launch, is too abrupt, introduction being most logically delineated by prominent activities specific to product launch.

To begin, the mechanism and timing of product launch needs to be chosen. The possibilities are enormous, and the following list is a whirlwind tour of general options.

- *Special events.* It is helpful for young companies to tie product launch to a large event, since they usually lack the resources or drawing power to host a glittering event themselves. Annual or semiannual national meetings (e.g., COMDEX for computationally related hardware and software) or other industry gatherings work well for most products, and an exhibit booth is the best launch mechanism. A low-budget approach is to hold private showings in a hotel suite.

- *Preannounced debut.* Another method is to hold a launch event of your own. Use early, broad publicity to indicate when and where the product will be shown, perhaps an airport hotel or other accessible location, the distinction being to avoid the link with a highly attended industry event and attendant competition. Rather, you hold an open house either alone or in association with some small industry event (a special workshop, symposium or other gathering) free of competition. This method can also salvage the introduction if a large industry event was missed.

- *Media blitz.* Events are not always important. Heavy, broad promotion of the product and its available date may be enough. Here it is best to offer a free trial period, to overcome the lack of hands-on demonstrations. This can be especially effective if you have a company sales force or an array of manufacturers' representatives in place, fully trained in product application and operation.

- *Mailings.* In an industry with known customers, an orchestrated series or perpetual stream of mailed pieces can be used. It is best to develop a set of fixed selling pieces like brochures and price sheets and at least one variable element such as a newsletter to establish a pattern of periodic mailings. By carefully avoiding excessive repetition, a program of this nature can run continuously.

- *Gradual ramp-up.* Another approach is slow, progressive ramping, which works best with expensive products and several key customers who have been nurtured during development. The few initial product placements provide revenue, buying time to gradually build further sales. Personal contact with the ultimate customer is paramount early on.

Many of the categories can be used in parallel, given the right product or industry and sufficient resources.

Once the product launch mechanism has been decided, you must properly prepare for the happening. Much general work has been done in the preparation stage, but here you must plan and execute all the fine details. We will not dwell on the issue due to project variability, but pull out all stops to ensure a smooth rollout—a lot is riding on the outcome.

Depending on the chosen approach, the grand opening itself may be either an event or an extended process. You should be thoroughly prepared relative to this checklist:

- *Administration.* All planning and modeling are in place, including backup plans, and all projected funding needs have been met or are committed to carry the project through introduction; management team is complete and adequate personnel and facilities are available for all relevant activities; comprehensive company controls are in place, including written SOPs, strategic management and planning, and an established QA function.

- *Technology.* Product design is complete, including all documentation; production setup is in early operation and with validation of initial product underway or recently completed; QC procedures and systems are in place and fully functional.

- *Marketing.* Market segmentation and customer profiles, derived from detailed marketing research, are in place for initial product; marketing program is fully designed and implemented for initial product; product imaging, positioning, promotion, pricing, and delivery are fully defined and in place for the inaugural product.

The outlined companywide posture provides a strong position of control. Even so, you cannot afford to relax, because surprises happen. Set up an early warning system among your group, composed of specified expectations quantified to the degree possible, and use the system to assess and fix plan deviations early. It takes discipline and diligence to keep the project on plan.

If something unravels, fall back on plan B. You certainly want to have a plan B in the wings, just in case. In fact, a plan C, even a plan D, is advisable, too, because making earth-shaking strategic decisions on the fly is treacherous. Develop the main and alternative plans in the relative calm of prelaunch planning. Take the alternative plans seriously, because you may have to use them. Management must decide the depth of contingency plans appropriate to their situation, but a good rule of thumb is to cover all major alternative product introduction outcomes considered at all likely (say, those with a greater-than-10% probability).

Eventually, the grand opening will be over. The next task is to evaluate where you stand relative to plan, carefully reassessing the entire financial model to create yet another version of that living document.

In principle, there is no stark contrast if you were forced to jump from plan A to plan B, C, or D during the grand opening, since the same assessment and model update must be created. Still, the new planning effort often must be more intense. A continuation of plan A may not involve much reevaluation of strategy, just rechecking and tuning. If an alternative plan was employed, however, strategy likely will need a keen reassessment. Of course, the tactics to implement each strategy should be rethought as well, whether or not the strategy has changed.

After the grand opening, the central mindset should be to establish sales momentum. If the grand opening created some product demand and perhaps sales, build on that base. If the results were below estimates, the original marketing approach must be supplanted or augmented. The available marketing approaches have not changed; they remain as outlined previously. So either continue the original grand opening approach, but harder and faster, or develop a new plan and implement it immediately.

Regardless of the approach, speed is essential to retain any acquired name and product recognition, since identity brings a new problem. You have appeared on the radar screen, and people are watching. If you suddenly disappear, the normal interpretation is a crash or other miscue. Realize that once you stumble it is far harder to build momentum a second time.

In summary, the varieties of approach open to different products and industries during product introduction are staggering. We can provide only an overview and shopping list of general methods. Literature may provide invaluable information about details, so check your local libraries, bookstores, Internet book sites, and other sources for appropriate references. University libraries often are superior to public ones. The Internet Web site amazon.com is a superb source for out-of-print books, even better than off-line sources, but barnesandnoble.com seems to have better depth of coverage for new technical books.

11.2 Monitoring of All Functions

The dynamics just do not get much higher than the introduction stage. When things are popping all around, it is hard to pay attention to plan, but it is imperative that you do so.

Administration's planning function is where activity and performance targets are found. Planning has laid out an intended trajectory, and all company functions are charged with maintaining operations on the intended path.

Another function of planning is to assess trajectory adherence. Central to the idea is an array of measures created throughout the company to monitor individual functions, analogous to manufacturing control charts as an early warning system but applied in parallel to diverse company operations. It is best to design a system with just enough sensors to give a sound snapshot of company performance; too many measures hamper analysis, while too few measures supply insufficient data.

Appropriate measures to assess performance are highly company dependent. Key variables include the nature of the product, company, and its state of evolution—parameters featuring so much variability that a fixed profile of measures is not possible. Instead, the following subsection presents a number of prototypic issues applicable, in turn, to each of the three company divisions.

11.2.1 Administration

First we address administration and the means to evaluate each of its several main functions.

- *Financial status.* This appraisal is best done using standard financial statements. An income (profit and loss) statement, balance sheet, and cash flow analysis are the most common statements employed. Your in-house expert or outside accounting firm can define statement content and any further reports of interest. All details are important, but prominent items of interest include revenue and its breakdown by source, direct sales costs, COGS, and general costs by company division. In the author's opinion, Intuit's Quickbooks® is the best accounting software for company use, at least through product introduction, but consult with accounting expertise for the final company choice.

- *Planning.* A means is required to plot actual against target performance. Useful measures might include costs, activity durations, adherence to schedules, and person-hours of effort. Regardless, be quantitative, to maximize objectivity. Planning tools like Microsoft Project® and Primavera's SureTrak Project Manager® can offer substantial benefit.

- *The plan and funding.* Closely allied with the planning assessment is ensuring that the business plan and the funding program are properly in place and on track.

- *QA.* First, ensure that SOPs exist for all company functions. Second, conduct scheduled and unscheduled audits to check actual versus SOP-specified performance. Third, design, install, and monitor quality costs (detailed in Chapter 13) and optimize them by adjusting SOPs.

- *Personnel.* Monitoring should be simple and direct. Attendance and general work quality (based on the results of periodic performance reviews) are useful, as are any patterns by work group or company division.

- *Facilities.* As with personnel, there should be simple checks to verify that the nature and arrangement of the facilities are adequate, the physical conditions and working environment are satisfactory, and any agreements and conditions relating to facilities possession are intact, proper, and being met.

Of the three company divisions, administration covers the broadest range of endeavors. Be alert to central problems that might have peripheral symptoms (on detecting a problem, check for ramifications elsewhere). Company efficiency is a must, but the company also needs good working relationships and proper rewards and respect for accomplishment.

11.2.2 Technology

Here we discuss the main measures of the function and performance of the technology division, the discussion being generic because of high intercompany variability.

- *Engineering.* Candidate measures of engineering (including R&D) performance include costs, human resources, task duration, schedule adherence, and a quantitative assessment of relevant general, verification, and validation design reviews.

- *Document control.* Key issues are ensuring that controls are in place and are working and QA audits providing the requisite evaluation data.

- *Production.* Reasonable criteria are often total assembly time per unit of product, product throughput, rejection rates, and time in testing.

- *Purchasing.* Typical measures are QA audits of actual costs versus estimated best case costs and a means to quantitatively track rejection rate and parts and materials availability by vendor.

- *Receiving and shipping.* Key indices might feature measures of residence time in receiving (prior to transfer to an in-house destination) or shipping (before shipment). Additional indicators are handling errors and the nature, prevalence, and cost of damage.

- *Inventory control.* Good benchmarks can include residence time of components and of finished product in inventory, total inventory value by item and class, handling errors, and the nature, prevalence, and cost of damage.

- *QC.* QC is best when inspections and tests are quantitative and charted. It is frequently useful to use control charts, to reveal and correct trends before product deviates from spec. Such data also can be statistically analyzed and assessed for variance. Other valuable measures are percentage acceptance of product and the amount, time, and cost of in-process and finished product nonconformance processing and rework.

- *Regulatory compliance.* Management should plot the progress of product through all initial approval and compliance maintenance processes.

The list shows that technology is assessed differently from administration. There are more technology functions to monitor, and the activities are also more concrete and measurable.

11.2.3 Marketing

Finally, we examine typical measures to track the marketing division and its performance over time:

- *Marketing research.* Research is hard to assess quantitatively. Marketing research must address the right questions, develop functionally accurate market segmentation and other descriptive tools, and design and implement effective programs and materials. However, assessment of effectiveness is largely subjective and therefore highly dependent on broad and deep perspective. Schedule adherence can be directly measured, but whether data are properly interpreted is a management judgment call. Short of independent control studies (usually impractical), data quality can best be ensured through proper management team marketing expertise accompanied by method assessment and improvement through long-term comparison of predictions versus reality.

- *Marketing design.* The marketing division conducts much design. Marketing research protocols are created, marketing programs are formulated and carried out, promotional material, presentations, and exhibits are fashioned and conducted, and so forth. Effectiveness indices can involve actual versus predicted sales, plotting forecasting accuracy, adherence to budgets and schedules, and measures of customer satisfaction (e.g., complaints, feedback from salespeople, questionnaires).

- *Development.* Marketing helps with product design during the original design effort proper and through postproduct introduction design tuning that results from customer feedback, but it is difficult to quantify the effectiveness. Even so, management should qualitatively evaluate the marketing division's development contributions over time relative to reasonable expectations.

- *Promotion.* Costs and resultant sales can be quantified, and marketing research can study perceptions and other indicators. Promotion's impact is usually large and immediate and deserves continuous close scrutiny relative to resulting sales.

- *Sales.* Sales should be segregated to evaluate distinct forms of promotion, sales mode, and any other pertinent variables. Actual versus predicted sales should be constantly analyzed to evaluate forecasting methods and promotion and to track general planning.

- *Customer service.* This function often exhibits sizable impact. It is important to design and implement the following elements: measures of the incidence and nature of incoming contacts (phone, fax, e-mail, and Internet logs); records of comments and complaints, with appropriate information forwarded to appropriate company functions; and tracking of order-processing time and the incidence of mistakes.

- *Shipped product.* Finally, shipping records should be designed for monitoring orders, manufacturing response, purchasing, quality activities, and inventory and packaging functions.

Marketing measures usually are intermediate in rigor between the more qualitative administration and the quite quantitative technology divisions.

11.3 Balancing Demand Versus Supply

It is crucial to orchestrate a balance between supply and demand. The company represents supply, the source of product, and your marketplace affects demand. Most companies are at the mercy of the marketplace, having little influence over ultimate demand, and it may seem that supply-demand balance is uncontrollable.

Actually, much can be done. True, short of a guaranteed contract, full control is not possible, but supply and demand can be balanced in a general way for many products and markets. The trick is to carefully orchestrate promotion to maintain demand at least within a manageable range.

Demand can be managed using deep industry knowledge and effective marketing research. Deep industry knowledge is best acquired in the form of a key management team member, a sterling instance of why it is necessary to build a truly powerful management team. Power means significant depth and breadth of knowledge in all relevant fields, as well as the field of the product. Turning to marketing research, the results of studies conducted during previous stages should define product need, who needs it, how they buy (the buying mechanism), and when. The when needs further discussion.

If you are in the post–grand opening period and questioning sales efficacy, comprehensive knowledge of potential customers is essential. The buyer attributes listed previously are key, but realize that buyers fall on a bell-shaped curve (Figure 11.1). When a new product is released, a few pioneer customers buy very early, but most people wait awhile to see initial reactions. There are also diehards in virtually any marketplace who buy very late if at all. During product launch, all revenue must come from pioneers, so if you can devise a way to spot potential pioneers, they can be barraged with a mailing program, telemarketing, and anything else that seems reasonable. Incidentally, this may also be just the thing to revive a stumbling start.

Early marketing research can be vital. The addressed questions are essential, no matter how things fare, and they become even more critical when things go awry.

Let's now analyze even deeper troubles. Say you did not try or were not able to obtain effective marketing research data. Assume further that you have not been able to attract marketing talent to the group either. These handicaps normally should lead to terminating plans for product launch. Still, let's presume you were compelled to launch anyway, due to a window of opportunity

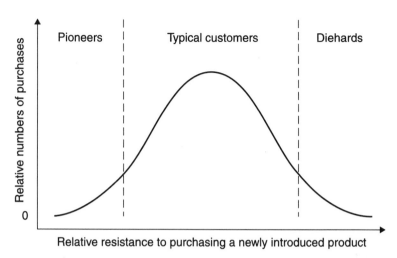

Figure 11.1 Graph of product purchase probability. Three subpopulations of potential product customers are qualitatively indicated.

or stockholder pressure. In essence, the project is out of control. You are suddenly compelled to do the marketing and other research that was neglected before, and more. Hit the libraries and the streets for needed data and expertise; also go to state agencies or the federal program SCORE (Service Corps of Retired Executives), which offers assistance for entrepreneurs. Analyze company data from initial sales activities and customer reactions. Pull out all the stops to develop a comprehensive rescue plan and then implement it.

11.4 Strategic Planning: An Update

In Section 11.2, we toured the three company divisions to see how company functions can be monitored, a process permitting quick corrective actions if the company strays from plan. Administration's planning function conducts the tracking and designs corrective actions.

We now go one step further, to examine companywide strategies that drive the functions. In subsequent sections, we also will see how companywide strategies affect each of the three company divisions. This is not an early warning system like the tracking function analyzed earlier. High-level strategy is the result of formal strategic management and planning.

Strategic planning and strategic management were introduced in Chapter 5. Strategic planning is the defining and implementing of long-term company objectives led by the planning function of administration. Keeping the ongoing company aligned with the strategic plan is called *strategic management*, carried out by the entire management team.

Strategic management and planning attend to nine essential factors, outlined in Chapter 5. Recall that, for consistency, we adopted elements from the superb book *Formulation and Implementation of Competitive Strategy*, by J. A. Pierce II and R. B. Robinson, Jr. In this chapter, we discuss how strategic management is applied (see Figure 11.2, which is adapted from the cited reference). We discuss this material here because the post–grand opening period is when the issues most likely will gain meaningful attention. Strategic methodologies should be introduced as early in the business development process as possible, but experience teaches that revenue must flow before internal systems gain reasonable priority. This attitude is especially notable in investors.

Figure 11.2 summarizes the strategic management process, adapted to present purposes. Basic terms briefly defined in Chapter 5 are explained below.

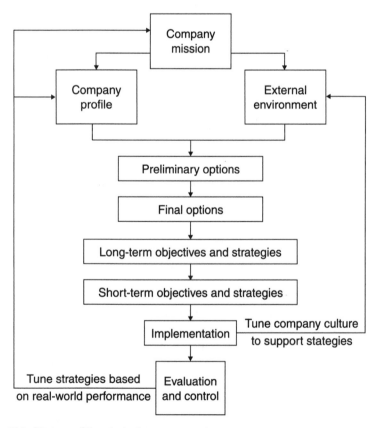

Figure 11.2 Diagram of the strategic management process.

- *Company mission.* The mission defines the company's product, market, and technological realms of emphasis, to show the purpose, philosophy, and goals of the board of directors and the management team.
- *Company profile.* At a given time, the profile lays out the quality and quantity of financial, human, and physical resources available to the company.
- *External environment.* There are two components to the external environment. First, the operating environment is composed of industry-related forces and conditions that affect strategic company moves. Examples include changes in user preference and actions by competitors, customers, suppliers, and creditors. Second, the remote environment concerns forces and conditions beyond the operating environment that determine the broad economic, political, social, and technological environment in which the industry exists. As cases in point, changes in credit (economic), regulations (political), demographics (social), or production methods (technological) can cause substantial upheaval.
- *Preliminary options.* Comparing the company profile to the external environment suggests a range of possible company opportunities.
- *Final options.* The possible opportunities are analyzed in light of the company mission to form an option list that is both possible and desirable. The winners are evaluated next to the long-term objectives and strategies to define one or more options expected to best achieve the company mission. The chosen options are called the *strategic choice.*
- *Long-term objectives and strategies.* Results sought over several years are termed long-term objectives. Examples include return on investment, profitability, competitive position, technological leadership, productivity, employee relations, public responsibility, and employee development. Objectives need to be specific, measurable, achievable, and consistent with other company objectives. The grand strategy is the complete plan of major actions by which the company intends to meet its long-term objectives.

  Every grand strategy will be unique. However, there are 12 recognized classes of grand strategy: innovation, market development, product development, concentration, vertical integration, horizontal integration, concentric diversification, joint venture, conglomerate diversification, retrenchment/turnaround, divestiture, and liquidation. The classes lay general boundary conditions on detailed strategy but leave room for individualized responses.
- *Short-term objectives and strategies.* This step translates the long-term objectives and grand strategy into specific annual objectives and functional

(departmental) strategies. The spirit of long- and short-term components is similar, but the annual objectives for the company and functional strategies for each department are more specific and must be coordinated across all functions to synergistically realize the company-wide annual objective. The annual objectives thus drive the functional strategies. Budgeting is coordinated along functional strategy lines.

- *Implementation.* This step lumps the more complicated breakdown of short-term objectives and strategies. The functional strategies must be translated into specific actions and targets for each function (department), in step with the company annual budget and SOPs. Company structure, leadership, and culture also must be properly considered, to effectively meld the plan into daily company life.

- *Evaluation and control.* Objectivity is the goal, but strategic plans encompass much subjectivity, so the real test comes only through actual practice. Two things can be done. First, observe the marketplace to assess plan efficacy and adjust the plan accordingly. Second, employ monitoring and control methods to ensure adherence to the strategic plan. In our recommended company structure, that is the job of administration's QA function.

11.5 Introduction Stage Strategic Issues

This section presents representative strategic objectives appropriate for each of the three company divisions. For convenience, the headings and presentation order for the following discourse on each company division match the discussion of monitoring and tracking in Section 11.2. All exemplary strategic objectives are defined in measurable, quantitative terms, so that achievement can be unambiguously assessed. Note that the company is assumed to reside in the introduction stage, so long-term objectives at this point extend 5 years into the future and therefore do not emphasize immediate introduction stage activities.

11.5.1 Strategic Issues in Technology

Strategic objectives of the technology division are the common denominator for the following list. The theme is generic illustrations of long-term (5-year) strategic objectives; the first item includes short-term strategic objectives, to illustrate the distinction. Of course, appropriate objectives for a given firm would be highly product, industry, company, and time specific.

- *Engineering.* Long term: conceptualize, develop, and design evolutionary products for the domestic business computer software market with self-evident functional utility. Short term: complete the e-mail accounting package design. Also, conceptualize and begin development of products defined by the marketing division in the first and third quarters, respectively.

- *Document control.* Upgrade document control so that average document creation time is reduced by one-third, and all unrestricted staff have network access to any documents that are current, archived, or undergoing change.

- *Production.* Increase production throughput by 30% and decrease nonconforming product and scrap by 20%, through plant modernization and automation and by increasing line worker competence through increased training.

- *Purchasing.* Reduce purchasing costs by 15%, through the creation of long-term vendor agreements, computer links with vendors, upgraded just-in-time programs, and other means.

- *Receiving and shipping.* Use opportunistic purchases of reconditioned packaging and bulk transport equipment, to improve person-day throughput by 20% and reduce damage costs by 10%.

- *Inventory control.* Improve general inventory control to reduce costs by 12% while maintaining throughput, average residence time, and damage at traditional levels. Devise a system to control nonconforming product, reducing mixups with general inventory to average no more than one incident per month.

- *Quality control.* Reduce the incidence of nonconforming product reaching production by one-third, while maintaining historical QA efficiency measures.

- *Regulatory compliance.* Tighten product regulatory compliance, reducing annual historical compliance agency inquiries or other regulatory actions by 18%.

These are just examples, structured more to show objective types rather than internal consistency. Of course, your company's strategic planning must accomplish both. Note that the objectives are measurable, to reduce subjectivity.

11.5.2 Strategic Issues in Marketing

Here we continue brief samples of representative long-term (5-year) strategic objectives, focused on the marketing division:

- *Marketing research.* Segment all existing markets and create multiple-parameter descriptions for every customer class as a hedge against changing demographics and also altered user preferences, expectations, and requirements. Also characterize at least two new markets for existing products and four new markets as candidates for product development.

- *Marketing design.* Through improved marketing research and marketing programs, lower marketing design costs from 3.2% to 2.7% of sales. Improve forecasting accuracy from a prediction error (actual versus predicted sales) of 8% to 4% of sales.

- *Development.* Use marketing research data to (1) develop two new markets for existing products and (2) pick the two best of four new candidate markets.

- *Promotion.* Working with marketing research, reduce the cost of promotion from 6.1% to 5.6% of sales, while achieving the increased sales defined as follows.

- *Sales.* Increase sales 20% in existing markets and create two new markets that augment annual revenue as follows: none in year 1; 5% in year 2; 12% in year 3; 21% in year 4; and 34% in year 5. Improve market segmentation analysis techniques so that current abilities to identify product demographic makeup are progressively improved 6% per year.

- *Customer service.* Improve average telephone waiting time from 3.0 minutes to 2.0 minutes and fax and e-mail response times from 33 and 25 minutes to 25 and 20 minutes, respectively. Reduce customer service order processing time from 17 to 14 minutes and the average order mistake incidence from 1/28 to 1/50.

- *Shipped product.* Reduce shipping residence and handling times from 3.2 days and 16 minutes to 2.7 days and 13 minutes, respectively.

As always, strategic objectives must be quantitative. Note that the above objectives can be accurately measured with little or no ambiguity.

11.5.3   Strategic Issues in Administration

This treatment provides examples of long-term strategic objectives tailored to individual functions of the administrative division.

- *Financial status.* Reduce the preparation time of monthly financial statements from 13 working days to 8. Introduce a system to provide selective financial data to each department on a need-to-know basis.

- *Planning.* Expand the planning unit to form a complete strategic planning function as defined by the director of quality and progressively introduce matching strategic management techniques to upper and middle management through formal training led by the director of quality. Measure effectiveness by monitoring annual sales dollars per employee for the company and by department, achieving at least a 15% improvement for each index.

- *The plan and funding.* Update the president- and board-approved business plan semiannually or more often if needed. Complete all planned funding efforts in full and on time, ensuring that funding sources meet board-defined specifications.

- *QA.* Enhance company controls (defined reporting structure, management meetings, SOPs, management performance reviews) by adding a complete QA system modeled after the intent and details of the FDA CGMP regulations by the end of the second quarter. For tracking, add a companywide quality cost monitoring program by the end of the second quarter. Reduce quality costs from the third quarter figure by 1% of sales per half-year period thereafter, throughout this 5-year strategic plan. Randomly conduct an unannounced audit of each SOP at least once every year and develop a corrective action plan for all deviations within 1 month of respective audit completion.

- *Personnel.* Reduce nonsalary payroll expenses from 28% to 20% by gradually increasing employee average monthly hours. Ensure that all personnel performance reviews are conducted on time and reprimand noncomplying managers. Increase the employee benefits package by 4% of base salaries or to the extent allowed using the savings but constrained by the personnel budget.

- *Facilities.* Solve existing city and OSHA problems regarding plant deficiencies and working conditions by the end of the second quarter. Improve the plant safety record from 1.2 to 0.9 incidents per 1,000 person-hours.

Note that administration must face the rigors of efficiency and cost control like everyone else.

We have now examined strategic methodologies at the company level and for each of the three functional divisions. Now we address actions to resolve problems revealed by the described monitoring.

11.6 Company Offense and Defense

The goal is to build a company that is as bulletproof as possible, administratively, technically, and in the marketing arena. By adhering to the recipe of this book, you will have amassed an array of offensive and defensive elements, which are reviewed here.

Your basic offensive arsenal includes the seven success factors first defined in Chapter 2. Each success factor is listed below, along with its inherent character and effectiveness at the introduction stage of business development.

- *Perspective.* By completing the company development recipe to this point, you and the company necessarily will have built powerful product, business, and industry insight. As you enter the marketplace and consolidate a respective position, you must continue the neverending process of strengthening the success factors that got you here, thereby building further perspective.

- *Planning.* Recognize that this activity built the company's present strength. Reevaluate planning methods, leveraging accrued experience to further optimize this paramount ongoing function.

- *Management.* Management achieved introduction and did so by planning and then making the plan happen. This grand triumph deserves accolades. Even so, reassess and, if necessary, alter the makeup of the management team to properly posture the company relative to future leadership needs.

- *Funding.* The success to date is testimony to effective funding and attention to all other success factors, since the funding would not have arrived without broad and deep project strength. Nevertheless, you should analyze past experience to further improve future funding efforts. In addition, it is prudent to begin investigating conventional business borrowing opportunities (factoring receivables, business loans), since the era of realistic borrowing is approaching.

- *Product.* As introduction wanes, significant data regarding product appeal, utility, acceptance, and quality should be known. Use such information to modify product design, manufacturing, and quality and to further enhance technology control and efficiency.

- *Marketing.* The efficacy of marketing should also be clear as introduction closes. Use feedback from ongoing marketing and sales to appraise and improve all programs and procedures.

- *Location.* By now, prior location decisions probably have revealed their merit. The company is also advancing, and it is critical to revisit location issues as resources and logic permit. Waiting accentuates old problems, may induce new ones, and sends the wrong message to vendors and customers.

Success factors change in relative importance and impact as the project advances. Funding requirements decline after introduction, as do product dynamics. But marketing rises to center stage, and location can become a major issue. Planning must continually evaluate the success factors and all other measures of company function, to constantly optimize the company mission, profitability, and functional objectives.

Having addressed offense, we shift focus to the fundamentals of defense. The book's recipe progressively introduces the following six interrelated control programs and systems, as company activities and complexity advance. The order of the items reflects the typical sequence of appearance in company planning and dynamics.

- *Management.* The management team was also listed as an offensive element. The defensive emphasis is on the flexibility and prowess conveyed by a strong management team for dealing with all present and future challenges. Management is listed first, since the project founder is in essence the original manager.

- *Planning and the business plan.* Planning was also highlighted as an offensive strength. The advantage in defense is mounting experience, which conveys increased leverage to foresee and implement successful plans and actions. Planning is commonly the second defensive element to appear and is monumental in impact.

- *SOPs.* After planning, the SOPs typically arise next. SOPs can be introduced gradually, as needed, during the hectic stages prior to introduction. The SOPs form a major implementation vehicle of planning, acting as a reference frame guiding every sort of decision and action. They should be expanded progressively in depth and breadth as the company and its planning become more advanced and sophisticated.

- *QA system.* We introduced the concept of a companywide QA system in Chapter 10, based on the CGMPs of the FDA. CGMPs are absolute requirements for medical manufacturers, but they carry a wealth of wisdom for every product company. The CGMPs were introduced in the discussion of the design stage as one practical example of when to advance to tighter control, but management should periodically reassess control during all stages and adjust the spectrum of control factors appropriately to maintain sound company C^3.

- *Strategic planning and management.* As the operation matures, we suggest that early C^3 be augmented to the level of strategic management and planning. This is not semantics. Strategic management and planning are a sophisticated form of long-term management and control that arose and became formalized after World War II. The largest companies thoroughly developed the system, and the ideas then progressively spread to others, sometimes in less rigorous form. By or soon after the introduction stage, a briskly progressing company should seriously practice strategic management and planning principles and practices.

- *Total quality management (TQM).* TQM is yet a newer concept, useful in augmenting strategic management and planning and the QA system. TQM is more an attitude and culture adjustment than a distinct control mechanism, but TQM is a modern recognized approach that provides major benefits for the company, its products, its industry and community image, and its employees. Key advantages are found in cost control, product quality, and working environment. TQM will be formally introduced later (in Chapter 13), a logical stage relative to practical company evolution.

Note the passage through advancing phases where control becomes progressively more sophisticated and complete. Each new element is not necessarily more complicated than those before, and all prior controls are retained, perhaps in modified form, as the next level of control is incorporated. Thus, the progression leads to broader and deeper defense against straying from plan, augmenting the company's growing defense prowess and making it better poised to engage the future than before.

11.7 Corrective Actions for All Functions

The offensive elements described in Section 11.6 (the seven success factors) collectively represent the major power, breadth, and depth of the company and its components. In turn, the described defensive elements together equal formal control elements embedded in operations to keep the company on plan and are thus primarily designed to identify problems and impose corrective actions.

Purely setting up controls is not enough, because the reasons for deviations must be found, fixes devised, and solutions installed. We define the process of problem identification, study, and characterization, followed by the design and imposition of verified and validated cures as *corrective action.*

Corrective action must permeate the very core of the company's defensive control systems. Strategic management and the planning function must see that the latter principle is fully applied. Fast and thorough corrective action can save money and preserve professional energy needed elsewhere.

Corrective action must be pervasive, each company division carrying out its own corrective actions. Top management and the special QA function of administration also must be strongly involved, highlighting problems to be addressed and following through to ensure proper resolution. The SOPs additionally provide potent tools that specify actions and behavior, a reference for activity. Corrective action is everyone's job, and all employees at every level should watch for inefficiencies and waste and alert the proper authorities. Furthermore, company operations and culture, written and unwritten, should encourage everyone to identify and resolve problems. The ultimate paradigm in this regard is TQM, the main topic of Chapter 13.

We are at the end of Part III, Production Design and Launch. We have seen how a truly professional project can gain control of product and business development and forge through to marketplace entry. Even the little guys can succeed, if they adequately attend to all the factors crucial to the process.

We now turn to Part IV, Building Long-Term Value, for a look at how to orchestrate the company's post-introduction future. Building long-term company value proves to be a valuable uniting theme to address virtually all other long-term concerns.

Part IV
Building Long-Term Value

Parts I, II, and III covered background, planning and funding, and a detailed analysis of start-up business development stages, respectively. Now we focus on designing and developing the company to optimize long-term value. Value conveys advantages and opportunities far beyond dollar worth and, most important, greater company flexibility. In turn, flexibility provides improved capacity to tap opportunities.

Part IV assumes the company has completed the introduction stage of business development, to enter the stabilization and growth stage. Chapter 12 is directed to strategic methodologies, revisited in light of advanced company status. The initial focus is on overall company strategic objectives. We then outline implementation of strategies at both the company and division levels commonly important after completion of the introduction stage. Marketing is particularly critical here and is given substantial extra attention. Finally, discourse steps successively through all three company divisions, covering key activities and concerns likely prominent in the stabilization and growth stage.

Chapter 13 examines company and function control in considerable depth. One last control system is introduced, TQM, to show how TQM consolidates modern thought on integrated management and company control. A related tool, installing and tracking quality costs, is introduced as an added monitoring technique augmenting control. In addition, analysis of company strategy and the management team is employed to show how control is structured in theory and implemented in practice. Control issues then are extended to assess employee relations and external professional company support. Finally, we examine how control contributes to success and building company value.

Chapter 14 closes the main text by summarizing and selectively extending the book's primary points. Included are reviews and perspective of company structure, success factors, the special success factors of planning, funding, and management, stages of business development, the four recommended company control systems, and broadened discussion of quality. The book's seven success factors are then analyzed in detail, looking back over the entire business development process, an exercise that shows the success factors change considerably from business development's beginning to its end. Next, the book's major points are reviewed a second time to examine practical aspects of how and why things go wrong. The discussion and the book finish by describing personal traits most probable for technical product entrepreneurs. That permits self-comparison and provides insight on important attributes key to assembling the requisite project team.

12

Stabilization and Growth

12.1 Company Objectives and Strategies

This chapter begins the book's last main segment, Part IV. The premise is that business development has progressed beyond the product introduction stage, that is, through the concept, R&D, design, preparation, and introduction stages, to arrive at stabilization and growth. The discussion turns from detailed product and business development mechanics to concerns of company endurance and building long-term value.

The subject interval, the stabilization and growth stage, begins upon completion of marketing division activities dominating the preceding stage, introduction. This milestone is taken to occur when product portrayal ceases to focus on the "new" or "just introduced" product condition, a more reliable demarcation than the extreme interproject variability of company financial criteria. The end of the introduction stage, identical with the advent of stabilization and growth, varies greatly, depending on product nature and promotion, but the overall introduction stage most commonly lasts weeks to months.

The stabilization and growth stage features many considerations and characteristics common across diverse products and markets. Several prominent issues will be addressed. The backdrop of strategic management and planning is employed, carrying forth the theme of Chapter 11.

12.1.1 Long-Term Objectives

We assume that strategic management and planning now represent the primary character of company direction and control. This chapter extends orientation and instruction relative to strategic planning and strategic management, those methods being of utmost importance to long-term success. Strategic considerations may change drastically stage by stage in a company's early history. Later, during stabilization and growth, company issues and strategic content also may undergo large alterations of form and substance but, one would hope, in a more sedate fashion.

Strategic practitioners realize that maximizing short-term profits hardly ever is best for achieving sustained company growth and profitability. Most strategic managers distribute a small amount of profit here but reinvest the majority of profits in company development. Most companies are not profitable when the stabilization and growth stage begins. In fact, young companies commonly delay dividends for some time, profits going to company consolidation, growth, and expansion until financial stability seems secure. We strongly advocate that approach. As a contrasting example, situations such as the phenomenal runup of Internet stocks in the late 1990s are far too extraordinary and unpredictable to form any basis for strategic planning.

Striving for sustained prosperity, strategic planners usually employ seven classes of long-term objectives.

- *Profitability.* Successful company operations in the long term depend on attaining and maintaining acceptable profits. Useful objectives most often are formulated as a percentage of revenues, earnings per outstanding share of stock, or return on equity.
- *Productivity.* Companies should continuously strive for increased productivity. Typical objectives employ measures of throughput, production costs or elements thereof as a percentage of sales, acceptance or defect rates, and adherence to schedule.
- *Competitive position.* Often the best company success indicator is relative marketplace position compared to competitors. While a poor index for companies new to the market, it can be phased in as measurable market share develops. Another good indicator at any market share is total revenue.
- *Employee development.* People value growth and career opportunities, and such benefits can lead to better productivity and less employee turnover. An employee development objective frequently is included in strategic planning, requiring one or more meaningful measures of employee development, such as duration of employment, advancement intervals, or education completed as an employee.
- *Employee relations.* It is prudent to seek good employee relations, even when unions or other outside forces are not a factor. Proactive steps that anticipate employee needs and expectations can depict manage-

ment favorably, and many believe that improves productivity and employee loyalty through perceived management interest in worker welfare. Examples of objectives include benefits, safety programs, day care for workers, flex hours, telecommuting arrangements, worker involvement in identified supervision and management processes, quality circles and other technical worker involvement programs, and stock option plans.

- *Technological leadership.* The company will either lead the marketplace or be a follower. Rapid assumption of leadership is tough if a company is just entering a mature market, but it is assured, at least temporarily, if company product creates a new market. Most situations lie between those two extremes. In any case, decide where in the continuum you are now and define, through realistic objectives, where you want to be in the future.

- *Public responsibility.* Businesses recognize responsibilities to society at large, as well as to customers. Many exceed minimal standards, building reputations for fairly priced products and as responsible corporate citizens. Illustrative objective topics relate to educational and charitable contributions, minority training, public or political activity, community welfare, and urban renewal.

Sound objectives are challenging but achievable, clearly stated, measurable, and suited to long-range company aims. Objectives also should align with (not contradict) general preferences and perceptions of those implementing them. Finally, objectives should be malleable to large or unforeseen changes in company environment or the marketplace.

The stabilization and growth stage is distinct, as revealed by a summary of typical strategic concerns. We have discussed seven strategic objectives, which define desired end results. Now we turn to key strategies, plans of action defining how those objectives are to be accomplished.

The following discussion treats each previously identified objective in turn. Each item first addresses general companywide objectives and then reviews candidate strategies often relevant in the aftermath of the introduction stage.

12.1.2 Profitability

Instant profitability is not reasonable, because the company has just finished the introduction stage (Figure 12.1). An excellent objective would be a target date for cost versus revenue break-even status, and additional objectives should define revenue (not profitability) target dates. Establish a dependable and growing revenue stream by reinvesting revenues to build company strength and

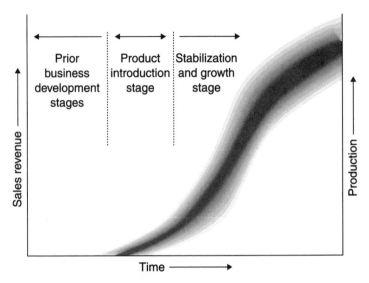

Figure 12.1 Qualitative revenue and production versus time. Axis scales have been omitted because of high interproject variability. Revenue and the underlying production remain at zero during initial product development and typically appear at product introduction at an initially low growth rate. After sustained promotion and sales have been active for some time, the sales rate progressively rises, represented by the portions of the curve with progressively increasing slope. Eventually, the slope magnitude usually stabilizes and turns slightly negative, as pronounced growth in sales magnitude continues, but the large and unsustainable early growth backs off to more reasonable long-term patterns. The curve is indicated as a broad band to highlight uncertainty.

value. First achieve a sound financial base (income statement, balance sheet, cash flow statement) before targeting any significant profits. For most companies, profitability becomes realistic several years after the introduction stage, but growth should be emphasized earlier. Reasonable company divisional strategies likely resemble the following selected items:

- *Administration.* Ensure that a full strategic planning system is continuously in place, covering both long-term (5-years) and short-term (1-year) periods. Lead a companywide SOP upgrade to be completed during the first year and conduct an SOP system review annually thereafter. Design and implement a TQM system over the first 2 plan years.

- *Marketing.* Contribute in step to administration's companywide SOP upgrade. Participate in ongoing top-level strategic planning and routinely implement within marketing. Achieve defined total revenue and market share targets for all plan periods. Define new products, for design by technology, necessary to achieve targeted revenue figures for each plan year.

- *Technology.* Contribute in step to administration's companywide SOP upgrade. Participate in ongoing top-level strategic planning and routinely implement within technology. Achieve defined design efforts on time, within budget, and within product cost limits.

12.1.3 Productivity

This long-term objective category carries major post–introduction stage weight in its own right and to offset residual startup inefficiencies. Indices of throughput, acceptance rates, defect percentages, and adherence to schedule should be defined. Progressively target reasonably dramatic improvements in all measures over the 5-year strategic plan (key examples are provided in the following material). Betterment rates should be particularly sharp at first, tailing off later as opportunities decline. Plausible divisional strategies are cited as follows:

- *Administration.* Upgrade SOPs during the first plan year to fully support the companywide strategic planning system and ensure comprehensive QA system existence over the same period. Develop and successively implement means to streamline company paperwork, reducing defined managerial examples by a defined percentage per year to be set by administration's strategic planning function. Meet specified facilities requirements and resources for work conduct for all planned phases of upcoming work.

- *Marketing.* Design or upgrade customer service to increase throughput and decrease error rates to staged, targeted values defined by the strategic planning function. Rigorously assess and design or upgrade programs as appropriate to progressively meet defined strategic planning objectives, for market research and customer analysis, program and promotional design, and new product concept development.

- *Technology.* Accomplish the following progressive system improvements as defined by the companywide strategic planning function: design or upgrade technical support to progressively increase throughput and decrease error rates; successively improve production by attaining staged, targeted, long-range values of defined measures; and annually reduce R&D and design cycle costs and duration, meeting progressively more stringent annual targets. Fully computerize the document control system, including change control, during the first plan year.

12.1.4   Competitive Position

This objective category is critical. The company's beginning position will vary enormously between projects, due to market nature and life cycle status. Virtually every situation, from embryonic to mature markets, demands concentrated efforts to properly position, promote, and sell new product. But equally crucial is product posturing early in the 5-year plan for objectives due later. As an illustration of approaches to assessing competitive position from the literature, Figure 12.2 briefly introduces the concept of perceptual maps.

Indices for early objectives should reference total revenue, with market share secondary. Midterm and late objectives should emphasize defined market share targets for each product and market, total revenue being important but secondary. These objectives might lead to the following illustrative divisional strategies:

- *Administration.* Define the coordinated, companywide means (strategies) for achieving the staged revenue and market share targets (objectives) for all intervals of the 5-year strategic plan. Identify and enroll the company in select trade and industry associations by the end of the second quarter and maintain active participation throughout the 5-year strategic plan.

- *Marketing.* Thoroughly evaluate present and future needs throughout the 5-year strategic plan and appropriately design and prepare to progressively implement enabling systems for the following marketing functions and across every unfolding company product (discussed in more detail in Section 12.3): product imaging, positioning, and pricing; product promotion and packaging; and distribution channels. Install a program for perpetual evaluation of the company marketplace environment to identify and plan products for new markets or segments fulfilling the 5-year strategic plan.

- *Technology.* Comprehensively assess the following activities and design or upgrade functions to meet defined progressive performance targets specified by strategic planning over the 5-year strategic plan: new product conceptualization, research, development, and design; manuals and documentation; technical support (scheduled person-hours, level of expertise); and reliability (defined in terms of MTBF).

12.1.5   Technological Leadership

For each relevant marketplace, the company must choose whether to lead or to follow relevant competitive products relative to product design approach,

(a) Two-dimensional grid

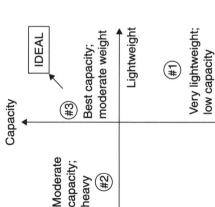

(b) Multidimensional table

Perceived heart pacemaker benefit scores*

Parameter	Competitor		
	#1	#2	#3
Size	5	4	3
Availability	4	5	4
Cost	2	3	5
Lifetime	4	5	3
Reliability (MTBF)	4	3	5
Quality	5	5	4
Standards compliance	5	5	5
Service and support	5	4	2
Reputation	5	4	4
Tenure in industry	4	5	2
Total score	43	43	37

*Score scale 0–5 (5 best).

Figure 12.2 Structured assessment of competitive position. Competitive position can be assessed by identifying all or a select subset of perceived customer benefits and systematically comparing your product to that of competitors. (a) Two significant dimensions for hypothetical batteries for consumer electronics products are depicted; the strengths of any given producer probably would be reflected in the promotional strategy. (b) A hypothetical example to show the treatment of multiple dimensions, the structure being tabular because of the difficulty of multidimensional graphical display. (Concepts adapted from G. L. Urban and J. R. Hauser, *Design and Marketing of New Products*, 2nd ed., McGraw-Hill, 1994, Chapter 8. Urban and Hauser describe the sophisticated process of competitive positioning relative to customer benefits.)

features, pricing, and a number of other variables. Objectives should be defined for intermediate and end point achievement. Distinct postures can be selected for different products or for the same product in different markets. Sample strategies related to technological leadership objectives follow.

- *Administration.* Progressively move both the administrative and marketing divisions (white collar functions) to the same modern facility with a contemporary look and setting by the end of the third quarter. Do the same for the technology division by eighth-quarter end, technology being relocated nearby the other divisions. Promote a uniform company technological posture, consistent with marketing strategies, in administration's dealings with industry, customers, competitors, vendors, and the community at large.

- *Marketing.* Emphasize the new white collar location in all appropriate promotional material by third-quarter end. Design all promotional material and product inserts to subtly convey technological leadership and an emphasis on quality and reliability, timed to begin with the move. Orchestrate the same points throughout the 5-year plan.

- *Technology.* Design future products with a contemporary and sturdy look and feel, conveying a modern appearance and robust performance. This must involve outside industrial designers, under an appropriately increased budget. Design manuals, through appearance and content, to exude quality, strength, and endurance. Progressively enhance the quality function over the plan interval, introducing TQM (see Section 13.2) beginning in the fifth quarter, to be fully in place by the end of the second plan year.

12.1.6   Employee Development and Relations

This objective class should receive moderate to high emphasis. Employees see young companies as having good and bad aspects. They are dynamic and often have interesting work with upward mobility, but they also are less stable and secure and are perceived as offering fewer benefits. Employee development is crucial to build loyalty, improve productivity, and control turnover. Two battles must be won. You must be perceived as developing employees and supporting their welfare to recruit attractive employees in the short term. Also, you must actually provide matching or superior development, benefits, and perks to retain loyal employees long term. Company strategic objectives should reflect those imperatives.

- *Administration.* Design an attractive employee program array for progressive implementation over the first 2 years of the 5-year plan. Continuously track employee sentiment through indices developed by planning. Install a semiannual personnel review system no later than second-quarter end, with stiff managerial reprisals for not meeting predefined schedules.

- *Marketing.* Design promotional material to subtly convey management's keen interest and active support of employee welfare and development. Develop an in-house newsletter, informing everyone about employee programs and major activities underway in the three company divisions. Implement both concepts by first-quarter end.

- *Technology.* Develop a quantitatively based merit system spanning all hourly technology division employees, tied to the personnel review system in a defined way, by fourth-quarter end.

12.1.7 Public Responsibility

Businesses recognize their responsibilities to society at large as well as to customers. Many companies go beyond minimal standards to instill reputations for fairly priced products and as being responsible corporate citizens. Illustrative objectives could relate to educational and charitable contributions, minority training, public or political activity, community welfare, and urban renewal. Possible strategies to implement those objectives follow.

- *Administration.* Orchestrate a local metropolitan area public image of strong company efforts supporting employee welfare and development by launching a permanent publicity campaign no later than second-quarter end. Announce and periodically remind the public that 3% of company profits will be donated to community economic development (introduce the program during the quarter of first expected profits).

- *Marketing.* Design and introduce by third-quarter end a free program to recycle all expended product returned to company facilities.

- *Technology.* Develop and implement a progressive program to reduce solid and liquid production waste by 5% per year throughout the 5-year strategic plan.

Note that many of the sample objectives and strategies listed in this section are very specific, while others are quite general. We employed some very general examples to introduce a few broad subjects for objectives and strategies. However, all your objectives and strategies must be very specific. Real objectives must be concrete, measurable results that are absolutely expected except in extraordinary circumstances. In turn, real strategies must be specified in quantitative terms and related to a defined achievement schedule. Otherwise, neither objectives nor strategies will serve your intended purposes.

The outlined objectives and strategies are merely suggestions. The intent is to show types of thinking useful in objective and strategy development. The specifics of your situation may differ substantially from the mainstream, high-growth, product-oriented company with market leadership opportunities assumed here. In any case, your strategic team must devise clear and concise objectives and strategies effective in controlling your environment, a substantial task that cannot be done in a few hours. Effective strategic planning and strategic management require considerable and continuous thought and effort, one you literally cannot afford slighting in any way. Pronounced dividends accrue from using sound strategic methodology, while neglecting such methods minimally negates the advantages and is usually worse. Customers have come to expect improved product and business quality associated with sound internal objectives, strategies, and control.

12.2   More About Marketing Strategies

Although we have covered all three functional company divisions, the stabilization and growth stage is particularly dependent on marketing. Success or failure rides on marketing's ability to build introduction stage sales into a progressively growing revenue stream (see Figure 12.1).

Marketing's pronounced importance argues for extra attention to its strategy. In fact, the following added marketing strategy perspective applies to all business development stages, subjects that also can be explored extensively in specialized books on marketing strategies (see Suggested Reading).

12.2.1   Ideology

For perspective, this book embodies one of four alternative marketing ideologies. We assume marketing research asks customers what they want, which is then designed and produced for them. Call this the *marketing approach*. Compare the *product approach*, which argues that product integrity overrides all other considerations, marketing being important but secondary. Or look at the *selling approach*, which states that strong selling and promotion are the primary way to move product. Finally, the *societal approach* extends the *marketing approach* to also emphasize long-term customer and societal welfare. Most new product companies do best, we believe, by following the *marketing approach*.

12.2.2   Alternative Plans

Marketing and the company are affected by many forces beyond immediate or even indirect control. Such factors arise from economic (recession), political (fiscal, monetary, and foreign trade policy), legal (laws, regulations), cultural (ethnic behaviors and beliefs), social (population and its demographics), scientific (new discoveries and understanding), technological (new practical materials and processes), and competitive (new players, new products) realms. While control is not possible, some degree of anticipation is, and management must be proactive, constantly scanning the environment for relevant influences. The companywide strategic planning function should include ongoing surveillance of the company environment for relevant influences. Strategic planning and management should then thoroughly analyze such reconnaissance to allow optimal adjustment of company objectives, strategies, and tactics. Such planning should include, as appropriate, the development of formal alternative or contingency plans accompanying the primary strategic plan (see Chapter 6).

12.2.3   Buyer Perceptions

Know your buyer. One way is to analyze buyers' perceptions, which fall into two classes:

- *Primary dimensions.* Potential customers view product advantages against perceived risks. Marketers design programs to emphasize the good points while offsetting or downplaying the bad.
- *Secondary dimensions.* This constellation includes how easily and how soon benefits will appear; simplicity of construction; ease and understanding of product and instructions for use; repair, maintenance, and return responsibilities and difficulties; compatibility of use with attitudes, opinions, and systems of belief; and product availability.

The listed items offer means to compare your product to the competition. Employ listed items that pertain to your product and add others deemed important for relative distinction. Then assign each item a weighting factor, to reflect item importance. Finally, on a scale of 1–10, score your product against the entire list and do the same for all relevant models of each competitor. Total scores reveal useful perspective on your competitive marketplace as seen through the dimension of buyer perceptions, information essential to validate product design and optimize promotional product differentiation at the stabilization and growth stage.

12.2.4   Participants in the Buying Process

Many people influence the buying of your product. You employ main features and many details that address end-user needs for product design. However, some end-users may have little to say about actual product purchase. Various entities may intervene between seller and end user, including a boss, a purchasing agent, a department head, or even a separate purchasing department. These intermediaries may be bound by purchasing specifications or rules laid down by upper management or a board of directors.

Ensure complete understanding of the buying process for your general marketplace and its individual components. While this is especially critical during the preparation stage or even earlier when this knowledge base is initially being assembled, it is critical to continue such activities through introduction, stabilization and growth, and beyond, to continually build perspective and contend with changing competitive positions and target environments. Be certain your promotional material and sales forms mesh properly with your knowledge base regarding customer structure and procedures.

12.2.5   Segmentation

Things frequently are grouped on the basis of common attributes. When talking about potential product purchasers, that process is called market segmentation. Companies employ segmentation to position products with a certain image and appeal. You first apply segmentation early in product development, as an aid to proper product design and in initial development of marketing concepts. On the basis of a customer knowledge base progressively developed over prior stages, segmentation becomes a prominent driving force for marketing during stabilization and growth (the introduction stage often has other central forces that shape marketing, such as central reliance on a major exhibit and leads obtained therefrom). Segmentation is implemented through coordinated promotion and sales, a process called target marketing.

We now discuss widely employed dimensions for market segmentation, including examples.

- *Geographic.* Those dimensions based on area, including neighborhood (e.g., zip code), city, state, region, country, and continent.
- *Demographic.* Those dimensions based on personal characteristics, like sex, age, gender preference, race, ethnicity, marital status, number of offspring, income, education, and so forth.
- *Usage related.* Those dimensions based on parameters of use, such as application, quantity, and timing.

- *Psychographic.* Those dimensions relying on attitudes, interests, and opinions (AIO) and including political, economic, technical, personal, occupational, sociocultural, religious, philosophical, aesthetic, and ethical/moral factors.

Analyze your product relative to those dimensions and subcomponents at all business development stages and identify market segments useful to your product sales. That may suggest particular ways to best gain efficient promotional and sales access to targeted groups. Market segmentation may help correlate your needs with extensive marketing information in the literature. Finally, the process may aid product design, revealing items or configurations targeted at unique groups. Segmentation becomes a mounting activity that should be a prominent ongoing formal marketing function by or before early stabilization and growth.

12.2.6 Classes of Marketing Strategy

Given an interpreted market segmentation, marketers can choose from three gross strategies for developing detailed strategic plans.

- *Concentrated marketing.* Introduction of one version of a product designed for a specific market segment.
- *Differentiated marketing.* Introduction of several product versions, each targeting a distinct market segment.
- *Undifferentiated marketing.* Introduction of a single, universal version that appeals to a basal common need or desire of an entire market.

The choice of gross strategy can exert significant influence on promotional and sales program design. As your initial product enters the market, the criticality of resultant revenue drives the always fundamental importance of marketing strategy to new heights. For instance, targeting one group in a highly segmented market may allow low-cost focused promotion such as direct mail. In contrast, direct mailings in an undifferentiated market usually would be impractical.

12.2.7 Detailed Strategy

Regardless of type, products have four general dimensions, often called the "four Ps" of the marketing mix.

- *Product* is an item offered to satisfy a customer need or desire. Besides customer focus, the marketer concentrates most on product differentiation, showing product distinction against competitive units.

- *Price* is key to strategy, because values above, equal to, or below various competing products can significantly influence sales.

- *Place* refers to distribution channels from producer to end user. Relevant functions include manufacturing, transportation, warehousing, wholesaling, and retailing.

- *Promotion* includes tools for communicating product attributes and corporate image to potential customers and end users. Some prominent options are mailings (including videos), telemarketing, direct sales, advertising, e-mail, product exhibits, collateral material, and public relations.

This survey can help you coordinate strategy, through direct illustrations of the four general dimensions of the marketing mix. Consult books on strategic marketing planning for detailed analyses.

A major purpose of meticulously building a management team with substantial breadth and depth tuned to your particular business is to have onboard expertise to carry out all the crucial activities discussed in this chapter and elsewhere. This introductory treatment of business development is limited to presenting the rationale for business formation, structure, and advancement, most detail being far beyond the scope of the presentation here.

12.2.8 Sales

You may read about different types of selling, but all selling ultimately relates to two models:

- *Consultative model.* This approach is customer focused. Salespeople carefully listen, identify both the customer's perceived and real problems, and identify solutions. Customers like the intense attention, but the salesperson's role is more passive and therefore provides less control.

- *AIDA model.* This approach depends on the power of persuasion. The salesperson assumes more control, but customers can feel coerced. The sale is carried out in an AIDA sequence of steps: *A* for getting their attention, *I* for acquiring interest, *D* for cultivating desire, and *A* for the call for action.

Neither model must be adopted totally, since aspects of each can be selected to build a composite presentation. Adjustments can accommodate personal style, cultural and industry norms, the customer, and time urgency. Some companies may need a relatively standard approach, to properly coordinate with a particular company image or customer base, but otherwise the choice depends on effectiveness, familiarity, and comfort.

12.3 Marketing Activities

This chapter emphasizes strategic planning and strategic management due to criticality. The ongoing activities that support and implement strategic planning and strategic management are pivotal as well. In that spirit, this and the next two sections address ongoing activities of the three functional company divisions.

The marketing division is addressed first because of the assumed state of company advancement. The introduction stage was recently finished, and the stabilization and growth stage is underway. Marketing is in the spotlight, the paramount concern for long-term company health being marketing performance. We now examine each marketing function in turn.

- *Marketing management.* This represents marketing division coordination, command, and control. Marketing management (the marketing part of overall strategic management) must monitor indices of division function defined by the company's strategic planning. Corrective action must be instigated immediately, given any deviations from company plan. Marketing management must also continually review and adjust strategic and tactical planning for the division itself. Finally, marketing must play its role in companywide strategic planning and management (composed of and conducted by management from all three company divisions).

- *Customer analysis.* This is marketing research directed at detailed characterizations of potential customers and markets, old and new. Also included are monitoring and evaluating the nature, needs, and satisfaction of existing company customers. The chief concerns during stabilization and growth are providing detailed reconnaissance for expanding existing markets and identifying and characterizing new markets.

- *Program design.* Designing various marketing programs draws on all other marketing activities in this list. Programs mean all forms of internal and external activity directed to marketing's supportive role involving upcoming products and also its direct product management of existing products. Included is marketing's responsibility to define new products and markets for technology's R&D. Additional issues include analyzing and tuning all internal marketing functions.

- *Imaging and positioning.* Marketing implements strategies posted by the company's strategic planning function. Strategies are implemented through promotional materials and activities, pricing, delivery channels, sales methods, and customer service, to accomplish coordinated product imaging and positioning in relevant marketplaces. One challenge is melding expanded and new products, impending or planned, into grand imaging and positioning design.

- *Promotion.* Promotion is a visible component of strategy implementation, often defining most product exposure of potential customers. Marketing must continually select the exact promotional method mix and intensity, across all products and markets, to accomplish company strategies.

- *Pricing.* This powerful component of the marketing plan can profoundly influence purchase decisions and may vary for distinct categories of customer, regions, or marketplaces. Those distinctions can be exploited to test different marketing or sales methods and conditions. Pricing must be constantly monitored and adjusted, to keep the marketing plan on track, being especially critical during stabilization and growth, when active positioning efforts related to company youth are still underway.

- *Delivery.* This means all product distribution channel activities (shipping, distributors or wholesalers, retailers, direct sales management, and so forth). The distribution channel system must be constantly monitored, to assess effectiveness relative to planned sales and to correct plan shortfalls or take advantage of new opportunities. The stabilization and growth stage can be quite dynamic, as new products and markets are pursued.

- *Sales and customer service.* These two primary marketing functions were discussed at some length in the marketing discussions in Sections 12.1 and 12.2. The management of these two functions typically reports to the marketing division head, but sales and customer service may be governed by the same person or distinct individuals. As the company advances through the introduction stage and on into stabilization and growth, the perpetual nature of sales and customer service criticality argues for progressively more concrete and formalized structure and procedures. Operation under the marketing umbrella is recommended to optimize coordination with all aspects of the product delivery chain.

12.4 Technology Activities

The fast-paced dynamics of the technology division characterize stabilization and growth. Stabilization does not mean stasis, exemplified by likely brisk production ramp-up with mounting sales and active R&D and product design through new product additions or expansion of existing product lines. The objectives and strategies of the technology division described in Section 12.1 are now reviewed in terms of stabilization and growth stage activities and concerns.

12.4.1 Technology Management

This represents technology division C^3. Technology management must monitor division functional measures defined by companywide strategic planning, and corrective action must be enacted immediately in response to any departures from plan. Technology management must also review and adjust division strategic and tactical planning, regularly addressing all departmental functions. Finally, technology must play its role as an integral part of companywide strategic planning and management.

The above considerations are true during all nonembryonic stages. The primary concern of the stabilization and growth stage is tight monitoring of the diverse development, design, and production functions that have come into play as initial product has flowed through the product development process for the first time. Technology must continue the development of future products in step with strategic planning needs while concurrently assessing and bolstering system design and implementation in light of accrued experience. The product environment has forever changed from an emphasis on development in earlier stages to a now perpetual dependence on parallel development, design, production, and product support.

12.4.2 R&D

R&D is an ongoing activity, taking direction from company strategic planning that defines required R&D pursuits. The output of R&D drives the technology division's detailed design and helps strategic planning better understand what is possible and plausible for future efforts. Critical R&D concerns in the stabilization and growth stage relate to deriving a systematic approach to what will be continuous R&D, improving efficiency of all R&D functions, and the capability of timely handling of new projects cast for study by strategic planning.

12.4.3 Design

Design extends R&D developments, takes direction and priorities from strategic planning, and specifies designs on which production runs. Design must conquer the conflicting stabilization and growth challenges of meeting schedules while achieving design specs, all in a continuous, dynamic, ongoing technical environment of necessity tightly coupled to a more comprehensive company strategic agenda, including broader product lines and market spectra.

12.4.4   Production

This function uses design documentation and perhaps ancillary design elements like fixtures or sample product to acquire raw materials and parts and to make product in volume. Chapter 13 introduces some modern refinements of production (and all other company functions), linking the process into overall company activity even tighter than historical standards. Classic challenges include holding purchasing, process, product, and QC to progressively tighter standards as time advances and various startup glitches are resolved.

12.4.5   Purchasing

Like production, purchasing must be closely controlled, or parts and materials will not be available when needed. Manufacturing's use of computerized ordering, inventory control, and production is virtually mandatory today. The management team must ensure suitable expertise for these demanding systems, and technology must continually optimize production control.

12.4.6   Inventory Control

Essentially, the same comments for purchasing apply to inventory control, because the two functions must be tightly coupled.

12.4.7   QC

QC must master four roles. First is the monitoring of results of key production steps, and second is final product test; both are tracked quantitatively to highlight trends and define thresholds for corrective action to avoid deviant product. Third, QC must inspect and control incoming (received) parts and materials until release to production, quantitative trends being monitored to better understand component lot characteristics and improve supplier control. Finally, QC must control address and disposition nonconforming product from suppliers, production, and returned finished goods, all nonconformity being analyzed to understand systematic or otherwise controllable conditions. QC must conduct these relevant functions during all business development stages, but the stabilization and growth stage is often distinguished by mounting of a real-world installed base, a source of potential problems (technical support, customer service) and perhaps returned goods (nonconforming product, warranty repairs). In addition, QC constantly must strive to tighten production and product control through improved capabilities of specification and assessment.

12.5 Administrative Activities

The stabilization and growth stage is no breather for the administrative division due to many challenges, including expansion, funding, resource provision and allocation, ensuring adequate facilities, consolidation of functions, and contending with mounting personnel count and mix. Here we highlight the activities and concerns of each administrative function.

12.5.1 General Management

General management embodies administration division C^3. Administration management must proctor division indices defined by the company's strategic planning, and corrective action must be carried out immediately to mitigate any divergence from company plan. Administrative management also must evaluate and adapt division strategic and tactical planning, assessing all departmental functions on a regular basis. Finally, administration must lead companywide strategic planning and management.

12.5.2 Finance

Finance analyzes past and present data to deploy company capital and monetary resources to maximum advantage. Day-to-day cash management is combined with long-term planning, through finance's participation in companywide strategic planning. Prime concerns during stabilization and growth usually shift from investor financing to business loans and bringing any contracted technology or marketing services in-house. Also, finance should provide guidance to companywide strategic management for increasing company value long term and periodically review accounting and budgeting systems for progressive optimization.

12.5.3 QA

Chapter 10 recommended the uncommon placement of a major quality function, QA, within administration. The benefits of a quality division are so strong that we now, in the stabilization and growth stage, suggest it as a standalone department, independent of both the administration division (where QA has resided until now) and the technology division (QC's prior placement). We envision the quality division reporting directly to the president, the reason for independence being objectivity. The quality division needs full authority to alter or shut down production and to manage other urgent problems detrimental to product quality or company health without undue pressure from multiple or biased superiors.

12.5.4 Personnel

The central concern is maximizing employee competency and motivation for improved productivity. Building employee development and employee relations was discussed in Section 12.1.

12.5.5 Facilities

Facilities include not only the physical buildings and their layout but equipment, furnishings, utilities, parking, and other related elements essential to productive company endeavor. The stabilization and growth stage can raise substantial challenges, since companies often outgrow physical resources due to rapid early growth but limited or delayed cash flow. Inattention to facilities can create severe constraints on the company's ability to meet demand created through aggressive marketing. The solution is to project and address facilities issues far ahead of need, providing time to arrange necessary resources and funding.

12.6 Business Valuation

This chapter first addressed companywide strategic objectives. Next, we visited strategies commonly important for the company and its three functional divisions in the post–introduction stage environment. In addition, marketing's extraordinary importance was recognized through extra background and strategic emphasis. Finally, we stepped through each company division, to study typical stabilization and growth stage activities and concerns.

A top objective during stabilization and growth should be building long-term company value. For best results, an understanding of how company value determinations are accomplished is essential. Business valuation is a complex standalone discipline, an excellent overview being Thomas Horn's *Business Valuation Manual* (see Suggested Reading), on which we draw.

Numerous distinct methods are used to value a business. For any situation, valuation is conducted using every method from the following roster of possibilities deemed applicable. Each chosen method is weighted to account for relative applicability, and the weighted values are summed and averaged to determine the sought valuation.

The described technique is robust, because all relevant valuation methods are employed, adjusted for relative importance. The list identifies those valuation methods that are reasonably reliable and potentially applicable to product-oriented, high-growth companies.

- *Ability-to-pay method.* Company value that can support a time purchase plan considered reasonable by typical business lending institutions;

- *Discounted cash flow method.* Sum of present value of adjusted annual returns for 5–10 years (exact interval based on particular industry and other factors);

- *Excess earnings method.* Market value of tangible assets plus a goodwill factor;

- *Economic value of assets method.* Market or liquidation (not replacement) value of tangible assets, plus allowance for sale value of intangibles;

- *Net worth per book method.* Total assets minus total liabilities, from the balance sheet;

- *Internal Revenue Service method.* Complicated IRS technique, based on net tangible assets plus excess over typical average earnings;

- *Comparable sales method.* Recent comparable acquisitions adjusted to allow for value-relevant differences;

- *Price-earnings ratio method.* Comparison to public companies with similar commercial activities, relative to stock price divided by net after-tax earnings per share, with due consideration to whether the subject company is also public;

- *Replacement cost method.* Sum of appraiser estimate of every tangible and intangible component required for total company recreation from scratch;

- *Secured loan value.* Value of the net tangible assets multiplied by the loan advance rate (a percentage derived from regional financial institutions);

- *Rule of thumb method.* Gross monthly or annual pretax or after-tax revenue multiplied by an accepted multiple for the subject industry.

These brief descriptions are not intended for direct use. For further explanations and application methods, consult the cited reference or other business valuation books. The goal here, consistent with the broad intent of this book, is to introduce and provide perspective on important considerations, many details being far beyond present scope.

Strategic planning should track company value using all applicable valuation methods and continuously groom the company to optimize valuation. The relevant methods, or at least their weightings, likely will change as business development advances. The purpose of tracking and grooming valuation is several-fold. First, interactions with investors or lenders can be strengthened with solid valuation backed by detailed data and rationale. Second, valuation data are an excellent company advancement index for stockholder reporting. Third, strategic planning's read on company progress is aided by reliable valuation analyses. Fourth, valuations are critical to support a private or public offering, one means to provide money to eventually retire early investors. Finally, the data are paramount to any discussions with other companies regarding mergers, acquisitions, or business sale.

The business has developed and matured considerably since inception. As the project has advanced, suggested early structure and organization have been progressively augmented with new functions and controls. We reiterate that all structure and controls should be designed and implemented as soon as practical during business development. However, real-world experience led us to delay presenting many functional and control elements until later stages, placing material near its likely point of practical application rather than at the academic ideal.

Chapter 13 addresses yet another form of available company control, TQM. TQM does not interfere with the control mechanisms previously discussed, like SOPs, a broad QA system, or strategic planning and management; those elements can remain in place, relatively undisturbed. TQM adds depth to overall control and instills improved employee development and relations. Thus, control remains at center stage.

13

Control: Making It Really Happen

13.1 Controlling Strategy

13.1.1 The Makeup of Control

Company control is essential at all times, but the business, industry, political, and social environments undergo incessant change. Even your primary industry, often seemingly stable to laypeople, frequently will be quite dynamic, so success in the business world is a moving target. Survival depends on diligent planning, skillful plan execution, and constant vigilance to effect rapid corrective planning.

Chapter 12 was almost completely devoted to strategic planning and its application to the company and its three divisions, administration, technology, and marketing. Strategic management's primary purpose is to effectively position and maneuver the company through a changing environment. There were times when periodic planning could launch company direction for significant unguided intervals, like a monotonous assembly line churning out identical desk calculators. Today, a better metaphor is an advanced assembly process in which each unit is custom built overnight to a unique user order (e.g., factory-direct computers). Modernized companies can carry out contingent strategic and tactical actions, achieving defined objectives that support a predefined mission in a constantly changing environment.

The major theme of Part IV is building long-term value. There is value in order and organization. Order means logical arrangement among a group's separate elements, while organization implies orchestrating a system into a prescribed, functional, structured whole. Order and organization are similar, but order is more like entity arrangement, while organization fits better with a functional process. In business, both order and organization are applied to achieve control.

Of course, order, organization, and control mean nothing without a sound product and marketing base. In turn, a great product and marketing base can be squandered easily. All these elements together are required to create value, while none is worth much alone.

In one view, though, control is at the pinnacle. Control has three meanings that, taken together, add up to unification and success. The context of the idea is C^3 (coordination, command, and control). First, control means to harmonize in a common action or effort the coordination component of C^3. Second, control implies authority to direct the course of action, the command component of C^3. Third is effecting restraints or checks, the control component. C^3 summarizes what the company must achieve internally.

For most industries, recent years have brought an acceleration of environmental change. Efficient strategy control and evaluation are paramount to ensure successful company direction through and adjustment to changing conditions. Previous discussion of company control is augmented here from a fresh vantage point to enhance understanding and perspective.

Strategic planning and management must measure performance, evaluate deviations from planned standards, and act to reinforce success and correct shortfalls or failure. Three fundamental dimensions are required:

- *Control.* Institute control systems based on performance standards coupled to annual objectives.

- *Monitoring.* Install efficient means to supervise performance and to identify and correct deviations.

- *Motivation.* Establish means to motivate all control, evaluation, and corrective action.

The process of controlling and evaluating strategy depends on effective implementation of all three elements.

13.1.2 Control Systems Linked to Annual Objectives

Two fundamental control paradigms exist for controlling and evaluating strategy implementation. First, *strategic control* is a top executive function that focuses on external forces and internal performance crucial to strategy success. Second, *operational control* is a concern of operating managers involving allocation and use of company human, physical, and financial resources.

Both levels of control, strategic and operational, share certain features (Table 13.1). For one, both must be linked to annual and long-term objectives.

Table 13.1
Features Shared by Strategic and Operational Control

Shared Features of Control	Strategic (Executive) Control	Operational (Function) Control
Links to annual and long-term objectives	Develop annual and long-term objectives that are coordinated across all company functions. Develop breakdowns of companywide objectives for every specific company function.	Managers of individual company functions must (1) study and understand overall company objectives and (2) develop specific strategies and tactics to implement objectives assigned to their specific function.
Clear, prioritized expectations with performance feedback	Define objectives quantitatively and prioritize them, for unambiguous assessment of achievement. Top management must provide detailed feedback on performance to function managers.	Individual function managers must translate companywide expectations and priorities into quantitative terms meaningful to their people and then routinely provide those people with constructive feedback on their performance.
Early warning system for deviations from plan	The company as a whole needs a formal system to track performance versus plan. Multiple-component warning systems are best, for cross-checks and redundancy. A companywide corrective action function is also critical to coordinate fixes.	Individual managers become the front-line defense by alerting executive management to deviations from plan and also serve as the front-line offense to implementing corrective action.

Recall that long-term objectives define company strategic goals for 5 years, while annual objectives dissect the broad 5-year goals into short-term targets. Effective control and evaluation means coupling short-term targets to specific performance standards, resource allocation guidelines, and strategic plan assumptions. The coupling defines the expected relation between strategic plans and requisite performance of company people and functional divisions.

Behavior is affected by the means used to measure it. That motivation basic leads to a second feature shared by strategic and operational control. People want to meet or exceed expectations, but those expectations often are not clear or reasonable. Effective control must plainly state expectations, provide ongoing performance feedback, and define priorities by relating expectations to strategic objectives. In that way, control and evaluation become positive motivational forces that encourage effective strategy implementation.

A third feature common to strategic and operational control is an early warning system of deviation from strategy or its implementation. The key is expression of objectives in terms of primary external assumptions, quantifiable performance standards, and resource allocation guides. Specific indices help people at all levels spot deviations from plan, so the proper authority can be alerted accordingly.

13.1.2.1   Strategic Control

Top management's marketplace understanding and company expectations form the basis for long-term objectives. Seven classes of long-term objectives were outlined in Section 12.1. Strategies appropriate to all three company divisions, administration, technology, and marketing, also were detailed in Sections 12.1 and 12.2. The advocated control and evaluation system bridges the gap between strategy development and implementation.

13.1.2.2   Operational Control

Previous references to control primarily have involved strategic control, but operating managers also need control and evaluation within their realm, governing allocation and use of company human, physical, and financial resources. Primary mechanisms for operational control are budgets and schedules.

Budgets

A budget is a resource allocation plan that helps coordinate company- and functional-level performance. Budgets do not exert direct control but rather allocate resources for specific company functions and represent standards against which action is measured.

Modern budgeting is used to define, confine, and monitor activity. Today, most firms employ a budgeting system rather than one large companywide budget. Figure 13.1 depicts a typical manufacturing company budgeting system; note the ties to the strategic planning and management process. Consistent with the figure, the control and evaluation system of most companies includes three budget types:

- *Revenue budgets.* A major company objective almost certainly will relate to revenue. Revenue performance is tracked through a revenue budget, a standard to compare sales and investment expectations (projections) against actual results. Revenue expectations come directly from the financial model of administration's planning function. Review intervals can be short, even daily, depending on what management deems meaningful.

- *Capital budgets.* The company allocates financial resources by this mechanism for major capital investment decisions. Capital budgeting covers the entire 5-year strategic planning window and carries substantial impact for young, high-growth manufacturing companies. Capital budgets set long-term strategy for each division's basic resources.

- *Expenditure budgets.* All company functions operate under an expenditure budget. Embryonic companies may combine sales, capital, and expenditures into one master budget, but separate sales, capital, and expenditure budgets will be advisable eventually, probably around the introduction stage, plus or minus one stage. Later, further budget breakdown may be prudent.

Budgeting is central to strategic management and planning and to requisite control and evaluation inherent therein.

Schedules

Timing is often key to strategy success, doing the right thing at the right time and place, so careful design and time strategy implementation usually are critical.

Scheduling is a planning and control tool for allocating resources and setting activity sequences, both being essential to successful strategy realization. Scheduling offers a mechanism to plan, implement, evaluate, and control elements of strategic planning and management.

The art and science of scheduling have become quite quantitative over recent decades. Formal techniques such as the critical path method (CPM) and the program evaluation and review technique (PERT) have become well developed and established. More recently, these techniques have been implemented in commercially available engineering software, exemplified by project management packages such as Microsoft's MS Project® and Primavera's SureTrac Project Manager®.

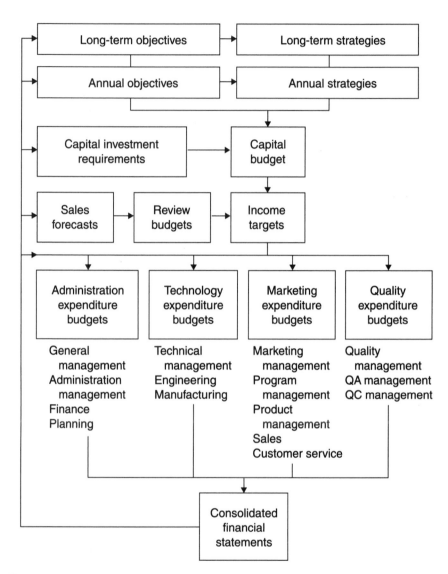

Figure 13.1 Typical budgeting system for a manufacturing company. The two long-term and two annual blocks at the top represent the long-term objectives and strategies and the short-term objectives and strategies blocks from Figure 11.2. The rest of the figure represents the budgetary components of Figure 11.2's implementation and evaluation and control blocks.

Like budgeting, scheduling finds major applications at both strategic (companywide) and operations (functional) company levels. As with other dimensions of strategic planning, budgeting and scheduling follow an iterative, company-level to functional-level and back to company-level cycle as planning asymptotically approaches final plan.

13.2 Total Quality Management

The ultracompetitive modern business environment is challenging companies as never before. Performance must be constantly increased by continuously improving processes, cutting costs, and increasing productivity, a truism for established companies large and small. As a result, the unprepared find it particularly hard to build a company, while the prepared enter a select group where prospects are vastly improved.

TQM centers on individuals and their companies, helping both employ total quality principles to understand customer requirements and to do things right the first time. TQM is a process and a way of thinking for conducting today's and tomorrow's business. In essentially every industry, TQM firms are emerging as tomorrow's leaders. If you plan on the long haul, plan on TQM.

TQM must flow from the top, being a management process and a set of disciplines coordinated to ensure that the company systematically meets or exceeds customer requirements. The company and all subordinate divisions and functions are deeply involved. All strategy and operations focus on customer needs, with high employee participation at all organizational levels. TQM companies act to systematically control, evaluate, and manage all functions at every level, to eliminate waste and to pursue continuous improvement.

TQM's goal is to deliver highest customer value at the lowest cost while achieving sustained company profit and stability. TQM can occur only with full management commitment to program design and implementation. Commitment requires building long-term relations with employees, suppliers, and customers and attending to quality beyond short-term profit. If you seek long-term success, there may be little choice, because TQM represents the future as envisioned by most who counsel. Customers have begun to experience the superior results and likely will not settle for less.

Listed are key characteristics that set TQM companies apart. The italicized entries are major TQM philosophies, contrasted against common pre-TQM practices.

- *Total customer service* versus less than total customer satisfaction accepted in-house to the extent tolerated by the customer base;

- *Long-term commitment* versus objectives and strategies emphasizing short-term profitability, not long-term commitment to customer interests;

- *Customer-driven philosophy* versus company mission with objectives and short- and long-term strategies being company-centric (self-serving) and customer consideration being recognized but secondary;

- *Continuous improvement* versus production at relatively high cost and waste considered a concomitant by-product of production activity;

- *Elimination of waste* versus production characterized by high scrap and rework, recognized as yet another natural result of production;

- *High quality and low cost* versus production that is relatively low quality and high cost;

- *Quality at the source* versus quality viewed as an attribute of inspection after the fact;

- *Leading people and measuring variance* versus people rated rather than cultivated and factory performance measured instead of product quality;

- *Cross-functional teams* versus parochial barriers between departments and other functional groups, resulting in poor interfunction communications and reduced overall company efficiency;

- *High employee participation* versus traditional top-down hierarchy that inhibits efficient interaction;

- *Multilevel communications* versus conventionally formal channels of top-down communication.

High-growth, product-oriented companies that insist on perpetuating traditional practices from the past have a threatened modern business existence. The manufacturing world is changing to favor the customer, and prudence argues following suit.

13.3 Implementing TQM

TQM provides impressive advantages for companies who adopt its principles, because TQM companies run better. Customers develop loyalty, attracted by company responsiveness to their needs. Internally, TQM companies develop high-performance, cross-functional teams, and a fertile environment for constructive improvement of process and action. Protocols are designed, data are collected, and analysis occurs for continuous process improvements. TQM companies invest in training and measure its value by assessing workplace impact. Employees like it, through improved self-expression, involvement, and job satisfaction, and suppliers and unions buy in too. The result is long-term productivity and product quality improvements at lower customer cost.

For TQM success in your company, it must become integral to the business. Either you are fully TQM or you are not TQM at all. TQM is an attitude for total commitment to customer satisfaction on every occasion, while introducing selective quality programs in an otherwise traditional business merely gives you the old business with improved quality, but not TQM. TQM mandates total top-down commitment, pervading every internal and external activity, requiring commitment, time, and discipline. TQM is like experience—there is only one way to get it.

The following list describes a model process for implementing TQM, the process being graphically depicted in Figure 13.2.

- *Management commitment.* Total top-down TQM management commitment is required, including sufficient implementation and sustaining resources.

- *Baseline assessment.* It is critical to audit the organization, to define the pre-TQM baseline. Often, an outside consultant is prudent for objectivity.

- *Implementation plan and resources.* After a baseline assessment analysis, management develops TQM objectives and strategies (the TQM plan) and allocates companywide resources to ensure due attention and commitment.

- *Implementation committee.* It is best to appoint a group to oversee TQM implementation. Retain the group long term for general TQM oversight.

- *Preparation.* Preparing to implement TQM or any change process must be finely adjusted and integrated beforehand. Develop strategy, assign roles, assemble resources, and ready the environment to accept altered methods.

- *Implementation.* TQM principles embody profound focus on customer requirements, people skills, measurement, proven quality programs, and customer service and satisfaction.

- *Evaluation.* Comprehensive management control and evaluation must track the TQM program, to assess and optimize plan achievement.

Once initiated, TQM is continuous. The model in Figure 13.2 shows TQM directed by the implementation group long term, although other implementations

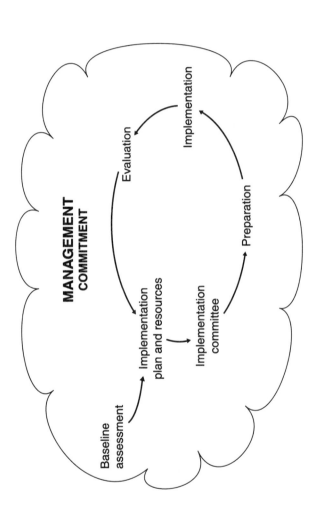

Figure 13.2 TQM implementation. The most critical element is total management commitment to the entire program. Baseline assessment occurs just once, to set the other recursive process in motion. Repeated cycles are required to fully implement and sustain TQM.

might suffice. Regardless, the process must be customer driven, have universal top-management support, be allocated proper resources, and have the authority to effect necessary change.

13.4 TQM and Strategic Management

The company already features strategic management and planning, SOPs, and a broad QA system. However, none of those pre-TQM systems inherently establishes TQM's main dictums: waste elimination, error-free work, and continuous improvement.

The pre-TQM systems are effective to a point, but they perpetuate existing cultures and mindsets. In industry historically, waste and errors were accepted as inherent to production, and company strategy was self-directed rather than focused on customer needs and satisfaction. The TQM ideal arose in Japan several decades ago (ironically being introduced by American consultants) and much later was adopted by major domestic companies. The most striking example is the U.S. automobile companies, which were forced to begin progressively adopting TQM principles by the commercial invasion of low-cost Japanese cars in the 1960s and 1970s. Now leading-edge product manufacturers have little choice, because many customers demand TQM.

TQM does not disturb strategic management and planning or other existing control systems but enhances those systems through a pervasive attitude change. Figure 13.3 portrays the resulting arrangement, a combination of all major control systems introduced in the book and summarized in the following list.

- *TQM.* TQM is shown enshrouding the whole company, to emphasize pervasive cultural and functional impact on every company activity.
- *QA.* QA is nearly all-encompassing too, as depicted in Figure 13.3. But QA is a procedural standard rather than a behavioral and attitude reference like TQM. The quality function (QA and QC together) should be relatively independent of both administration and technology authority, to preserve objective decisionmaking and action.
- *Strategic planning and management.* A separate strategic planning function collects companywide reconnaissance and acts through executive management (perhaps the same top-management individuals) to effect control and corrective action. Executive management is given a strategic tag to highlight the latter point.

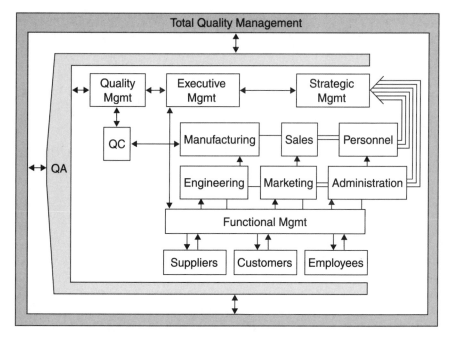

Figure 13.3 Complete array of company control systems: SOPs (not shown directly but promulgated by executive management), QA, strategic management, and TQM.

- *SOPs.* SOPs define specific protocols or guidelines for every key company function. SOPs operate at the division or department level, being enforced by respective group directors.

TQM adds considerable clout to an already sound system. It is also visible and makes a substantial contribution to company performance and image.

13.5   Quality Costs

One management method for refining companywide quality is installation of a formal program for measuring quality costs. Quality cost techniques systematically measure, report, and track the effectiveness and efficiency of relevant subelements of all company functions that contribute directly or indirectly to product quality. The spectrum of relevant activities is diverse indeed, as we will show.

Monitoring specific activities related to quality produces data valuable in managerial control, the main purpose of quality. Additionally, the formal system for monitoring quality costs serves to emphasize that quality is important, and the surprising breadth of activities monitored (to be detailed shortly) helps convey how pervasive true attention to quality must be. The specific aspects of diverse company functions amenable to quality cost measurements are described in the following discussion under four recognized quality cost categories.

13.5.1 Prevention Costs

The focus here is monitoring a specific subset of quality costs to improve control of product development, design, and manufacture.

- *Product development and design.* These activities are affected by quality personnel during technical development and design, such as general quality function efforts; quality participation in design reviews, qualification tests, and field trials; and all design verification and validation.
- *Marketing.* Included here are user need surveys, sales contract reviews, and quality-applicable proportions of marketing research, customer perception, and satisfaction studies.
- *Purchasing.* Functions include reviews and ratings of suppliers, quality planning with suppliers, and design and review of technical purchase specifications.
- *Operations.* Significant factors include quality planning and review, process validation, operator training, and the quality portion of operations design and development.
- *Quality administration.* Key elements are administrative salaries and expenses, training, quality planning, the total QA program, TQM and SOP management, and quality contributions to executive management.

Substantial issues arise before any product is produced, and production adds several more. Notice that the collective list of cited functions involves only a select set of total product development, design, and manufacture activities, but that the subset of focus (combined with added subsets described in immediately following material) precisely singles out those activities that contribute to a global picture of product quality planning, design, and implementation.

13.5.2 Appraisal Costs

Quality functions associated with or affecting direct product or component assessment reside in this cost class.

- *Purchasing.* Main elements involve receiving inspection activities and equipment, supplier certification, and component qualification activities.

- *Operations.* Key costs include process control activities and equipment, certification by standards and regulatory groups, ongoing operator training, production equipment repair and maintenance, and QC activities and equipment.

- *External to company.* Chief issues are technical evaluation of product in the field, marketing research on user preferences, and appraisal of spare parts and warranty activities.

Notice that costs are substantial just to observe product manufacture and normal use, the central-most aspects of quality manifestation.

13.5.3   Internal Failure Costs

This cost class arises from failed supplier or in-house activities that affect product manufacture.

- *Product development and design.* Prime items are corrective action, rework or scrap costs due to design changes, and ongoing engineering and quality interactions with manufacturing.

- *Purchasing.* Key activities include purchased material reject and replacement costs, rework of received material, supplier corrective action, and miscellaneous materials loss.

- *Operations.* Main issues are managing and dispositioning nonconforming product, problem and failure analysis and corrective action, reinspection, and excessive labor expenses.

Applying quality principles and practices here can usually produce substantial savings.

13.5.4   External Failure Costs

This cost family represents quality-dependent incidents in which failure occurs or is found after product manufacture.

- *Customer related.* Chief functions are customer complaint investigation and resolution, warranty costs, and customer goodwill.

- *Product related.* Main elements include disposition of returned goods (repair or replace), retrofit expenses, and recall costs.

- *Liability costs.* Significant factors involve insurance and direct company expenses.

- *Miscellaneous costs.* This assortment encompasses product- and operations-dependent penalties for regulatory or other infractions.

Especially with high warranty costs and customer goodwill, external failure costs can be a productive cost control target.

Quality costs seldom are heralded, but they represent substantial opportunities for insight and control. Insight comes from better understanding of expenditure sources, while control is improved through finding and resolving problems. Historically, quality costs represented up to a staggering one-third of total product costs. Worse, for those same companies, failure costs alone represented up to two-thirds of the quality costs. That is inexcusable, given modern quality wisdom and practices. Companies that tolerate such inefficiency and waste are out of control and highly vulnerable to savvy competitive forces.

13.6 Control and the Management Team

Doing the best job and achieving proper control requires the right top people. That assertion is now explained, along with how to acquire and keep top people, from the current post-introduction perspective.

Chapter 5 discussed building the management team, a paramount activity. We have seen the overriding importance of perspective, planning, the business plan, and funding to project and company success. Superior performance on all listed elements is best ensured through the makeup and dedication of the management team.

Managerial effectiveness is hard to generalize, since many skills are subjective and ends can be met in many ways. Commonalities exist, though, and the following list includes examples. Insist on solid tangible evidence on every attribute from business and personal references.

- *Leadership.* Good leaders personally show their people company objectives and strategy and why they point to success. Leaders also adhere to plan by ensuring strategy-congruent subordinate plans and activities. Finally, leaders emphasize qualities that distinguish the company from competitors and celebrate triumphs with their people. In good times or bad, leaders project a clear vision of future company prosperity.

- *Administrative skills.* Leadership is future oriented, while administration (management) focuses on getting today's work done. Administrative skills implement company and division strategic plans, through tactics and lesser order strategies applicable within and between individual company functions.

- *Motivation.* The management team must constantly motivate the other employees. People can be motivated by appeal, incentive, coercion, or force. Top managers create a culture in which appeal and incentive are the tools of verbal and written exchange, without resorting to threatening backups.

- *Strategic skills.* Perspective, planning, monitoring, and corrective action are the methods of strategic management. At least one or two team members must have real-world strategic management experience, with the rest capable, prepared, motivated, and fully behind the process.

- *Funding strength.* Funding typically is the most critical and difficult success factor. Evaluate every potential management member against the following standards: (1) direct production of funding candidates; (2) organization or group connections that can refer potential investors; and (3) attributes impressive to potential investors. Every team member must shine relative to the third item, and the team should collectively provide some strength relative to the first and second items.

- *People skills.* These skills include diplomacy, charm, efficient communications, projected respect, and fairness with and across individuals. These skills can materialize in countless ways. Face-to-face assessment, plus detailed discussions with professional and lay superiors, peers, subordinates, and others, is mandatory.

- *Credentials.* For a high-growth startup, impressive professional credentials across management are essential. The most significant reason is funding, since investors see little else. Management credentials also are important elsewhere, including employee recruitment and interaction with the external world.

- *Expertise.* Collectively, the management team must strongly cover the seven success factors, and each team member also must be strong in at least one success factor. Complete the management team as early as possible, and during times of special need in the interim, use consultants to plug any team gaps until the in-house team is intact.

- *Respect.* We have seen otherwise sound people whom employees do not respect. Respect resembles people skills but includes sincere regard and value for others, above, beside, and below in rank. Respect is a two-way street.

Every person considered for the management team must individually stand high against this collective attribute list.

13.7 Control, Employees, and Support

Not only the management team is critical to control, because the caliber of people working for or with the company can have tremendous impact. We review three human resource classes paramount to high-growth, product-oriented company control.

13.7.1 The Right Employees

A superior work force provides significant insurance for sustained control, and the best path to a top work force is the hiring of top people. The standard retort is that time never permits hiring people carefully, but strategically (long term) that viewpoint is backward. Actually, hiring people with low productivity, attitude problems, or lacking requisite skill costs the company dearly. Consider how such people would drive up quality costs alone. We recommend allocating proper time and other resources to hire correctly up front.

Hiring discipline is paramount, while shortcuts cause mistakes. These seemingly trivial points underlie most trouble. Do the homework; do not depend on instinct. Hiring should minimally include writing a detailed job description; an appropriate applicant search; building a realistic quantity of applicants; two successive rounds of resume sorting to build a controlled candidate list; a telephone interview with second-sort survivors and their personal and professional references; an initial in-person meeting with phone interview picks; second-round in-person meetings with survivors, including sessions with relevant staff members; and, if appropriate, the hiring act. Finally, should none of the candidates prove suitable, start over. The described process applies to essentially all hiring short of the management team, a highly special process detailed in Chapter 5.

The candidate search mechanism is also significant. Typical choices include help-wanted advertising; internal referrals; college, community college, and trade school placement offices; professional societies for particular specialties; professional publications; trade associations; outplacement offices of major manufacturers; and special job listing sites on the Internet (e.g., http://www.adamsjobbank.com). Sometimes employment contractors (temporary help) and employee leasing concerns are helpful, an approach amenable to taking some people on trial for possible later hiring. Employment agencies or unemployment offices usually are a poor choice, but consider the possibility.

Effective employee management can add to company value. Employee loyalty and longevity are improved through good vertical communications channels. Also key is effective feedback to let employees know where they stand, apprise them of weak points that need improvement, and recognize their strong points. It is best to diligently conduct employee performance reviews on an announced schedule. Performance reviews must be uniform across company divisions and functions, and managers must take them seriously, since performance reviews that are late, hurried, or otherwise slighted send employees a message of disrespect and indifference.

It is also imperative to retain good people once they are on board. A nonexhaustive list of considerations includes salaries, benefits, working hours, company culture and working environment, bonus plans, educational assistance, profit sharing or stock-option plans, retirement plan, and day care. Convention regarding those issues differs by region, state, locale, and industry and also varies significantly over time. Study your situation and properly address the four company-control systems outlined in Section 13.4. That should allow offering superior pay and other job attributes while maintaining a strong competitive industry position. Do not forget location, a success factor that can substantially affect company attractiveness.

A company with effective management and appreciative employees is probably under control. Given good business planning as well, the odds of success are greatly enhanced.

13.7.2   The Right Consultants

It is paramount to build a management team that effectively covers all important company strengths. Still, it is difficult to always have a complete team. For one, it takes time to start the project and create structure, and the project likely ramps up, so scheduled team slots appear spread over considerable time. In addition, an appropriate person may not arrive when needed, and filled slots suddenly may become vacant, time being required to replace key people.

But management team roles are so necessary that alternative plans become essential. The best backups for temporary or extended team shortfalls are consultants. In fact, we also suggest them for suitable nonmanagerial tasks rather than hiring employees. Consultants can avoid long-term commitments, resolve temporary management team gaps, and reduce costs of irregular and lower-level activities. Consultation also permits evaluating prospective employees on limited projects for possible future hire.

Retaining the right consultants at the right time can preserve tight schedules and save money. This excellent contingency measure can be useful in myriad situations, from the executive level on down.

13.7.3   The Right Professionals

Develop relationships with key professionals (legal, accounting, financial, and insurance) well before product introduction. Some will not be needed until later, but early rapport allows professionals to develop company and team familiarity and to respond rapidly should acute problems arise.

Professionals also may prove of invaluable funding help, as discussed in Chapter 10. Aligning with major professional firms may access contacts or clients who are potential investors. While top-drawer services may cost more, potential funding help could be priceless, so make sound choices of professionals and stick with them. Long-term relationships provide continuity and ongoing familiarity.

- *Legal professionals.* Unless conducted by a management team member, company incorporation may be the first professional need encountered. Lease agreements for space, patent and copyright work, and legal counsel on funding processes and stock purchase agreements also are plausible needs prior to product launch.

- *Accounting professionals.* Soon after project startup, accounting responsibilities likely arise for a management team member or outside professional. Key issues include federal and state filings, payroll mechanisms and filings, and any local taxes. Other relevant assistance includes design of a recordkeeping system, general tax counseling, and software package selection and assistance. For payroll, related taxes, and other human resources matters, retaining an outside payroll management firm may prove cost effective. Volume-based efficiency typically translates into reasonable rates.

- *Financial professionals.* A company financial milestone that occurs some time beyond the introduction stage is achieving company commercial loan status. Before this point, conventional business bank loans generally are not possible, since standard borrowing criteria (threshold cash flow, profitability) cannot be met. But past the threshold, a new funding avenue arises. A current business plan and an investor-style company presentation is still needed, and bank officials will also require historical and current standard financial statements. Management team strength also receives considerable weight, and loan issuance will depend as much on rapport and confidence as on financial data and business plan.

   Prior to achieving commercial loan status but after product launch, financing of receivables will likely prove possible to partially offset the delay between product shipment and receipt of respective payment. Such financing is expensive, so management must carefully weigh costs and risks against benefits.

- *Insurance professionals.* Probably the last required business professional class will represent insurance. Usually insurance is not needed before product field trails around the preparation stage or product field exposure beginning with the introduction stage. The main concerns are conventional business insurance for fire, theft, and other standard features and product liability insurance when trial or production units reach outside users. A buy-sell agreement also may exist between major players and investors, and it is often prudent to back up buy-sell agreements with life insurance on relevant signatories to fund buy-sell transactions if ever triggered.

- *Technical professionals.* The one nonbusiness consultant category of importance here is technical professionals. While product development and marketing consultants have been referenced elsewhere, the emphasis here is help with the central issues of this chapter, company control and our recommendations for interlaced quality and monitoring systems. There may be times when design, upgrading, or corrective action issues regarding company control systems (SOPs, strategic management and planning, QA, TQM) would benefit greatly from the objectivity and extra expertise of outside consultation. ISO 9000 production control issues also may become paramount, due to unique customer or regulatory requirements. Even though ISO 9000 concerns largely parallel the FDA CGMP guidelines emphasized in this book, external help may prove invaluable to address any singular ISO 9000 criterion that may become applicable in selected situations.

13.8   Control, Success, and Company Value

Given a project with sound perspective, planning, management, funding, products, marketing, and location, a company is in admirable shape. In addition to those seven success factors, a strong need exists for control, the central theme of this chapter, to keep the company on track.

Control is not viewed separately, like an eighth success factor. Control is composite, a result of perspective, planning, and management acting in concert.

For us, control arises by management's parallel use of the four control systems summarized in Section 13.4, strategic planning and management, SOPs, QA, and TQM. The four control systems form checks and balances that sustain the plan from different vantage points. The four-system combination provides powerful redundancy, ensuring plan adherence to any practical level desired. Of course, too much control is as bad as not enough. Practical management team expertise, common sense, and watching for overcontrol or undercontrol are tools of judicious direction.

Progressive installation of the four described control systems provides potent insurance protecting company investment and value. Control breeds success, and success breeds value.

We believe the best parameter to guide and motivate ongoing strategic choice is long-term company value. A value-coordinated mindset will almost always benefit other success measures too, such as company sales, profits, quality cost control, and so forth. Also use company value to cross-check more traditional considerations under ongoing strategic planning and implementation.

Success cannot be guaranteed, but the approach proposed here would be expected to bolster prospects substantially.

14

Final Overview

14.1 Major Points in Review

Our journey has surveyed business development, from concept to a self-sufficient growth company and from overview to fine detail. Perspective is now in order.

We began with a primal flash, a concept of product or user need orchestrated into verified product and business concepts. From there, the real product and business were designed and built. Let's review several significant points.

14.1.1 Company Structure

Company structure bears tremendous impact. While, in theory, any legal business entity will do, building a successful, product-oriented high-growth business places important constraints on organizational structure. Probably the biggest issue is funding, an essential activity that is always perilous, especially for entities that are not incorporated. Investors are familiar with corporations, understand how they work, how each stakeholder relates to the whole, and how ownership can be sold. Partnerships have never been attractive, because liability passes to partners just like a sole proprietorship. The limited liability company (LLC) is better, having liability protection equivalent to corporations, but the LLC structure is new and untested. Investor discomfort due to unfamiliarity, unknown ownership security, and unclear exit mechanisms makes funding difficult.

Entrepreneurial funding experience reveals that any negatives create serious barriers. View investors as looking hard for reasons to say no, so do not give them any.

14.1.2   Success Factors

Chapter 2 examined the seven success factors (perspective, planning, management, funding, products, marketing, and location). Properly addressed, these factors are common to diverse successful projects and represent virtually mandatory core elements for project structure, content, objectives, and guidance. Position a project for success by addressing every success factor to the greatest extent practical and logical.

We view the seven success factors as the most fundamental underpinning of winning business development. Later, in Section 14.2, all success factors are reviewed in retrospect, looking back over the project from the stabilization and growth stage.

14.1.3   Planning, the Business Plan, Funding, and Management

We must specially highlight four of the seven success factors. All the success factors are essential and call for concentrated focus. Still, planning, the business plan, funding, and management, working in concert, dominate significance and demands for excellence. They are the most visible expressions of successful management's three cornerstones: competence, motivation, and effective interpersonal dynamics.

On initial project exposure, outside investors and professionals develop a distinctive impression of the company and its management. The infamous "first impression" forms quickly on initial exposure and is difficult to alter later. Success requires routinely inducing impelling first impressions for both company and team, an achievement that depends on all seven success factors but especially the four highlighted here.

14.1.4   Stages of Business Development

The notion of business development stages was introduced in Chapter 1. Of course, the real process is a continuum, so staging is arbitrary but useful. Many business development factors change as the project and the business advance, and staging illuminates the special issues that frame their boundaries.

14.1.5   Four Control Systems

The organization of this book fundamentally followed company development over time. As discussion advanced, four independent control systems were progressively described that collectively act to keep activities on plan. The following list summarizes the discussion of the major control systems.

- *TQM.* TQM was first mentioned in Section 12.1, treated in perspective in Section 12.6, and detailed in Sections 13.2 to 13.4.

- *QA.* QA was introduced in Chapter 5, detailed in Section 10.2, and received important elaboration in Sections 12.5 and 13.4.

- *Strategic management and planning.* This methodology first appeared in Chapter 5, received a comprehensive update in Section 11.4, and underwent how-to elaborations in Sections 12.1, 12.2, 13.1, and 13.4.

- *SOPs.* SOPs were first discussed in Chapter 5, dissected at great length in Section 10.2, and placed in overall control system perspective in Section 13.4.

Each control system is potent and capable of independent function, but real strength comes from redundancy due to system overlap, distinct system perspectives, and perpetual effect realized through ongoing use.

14.1.6 Quality

This book emphasizes quality, major highlights including CGMP (largely analogous to ISO 9000) in Chapters 9 and 10, the case for a separate quality function in Chapter 12, and previous review of the four control systems. Quality and control go hand in hand, quality being a prominent enabling mechanism for control, the major topic of Chapter 13.

14.2 Success Factors in Review

The seven success factors described in Chapter 2 were used as reference frames for much discussion elsewhere. We now retrospectively analyze each success factor, availing ourselves of stabilization and growth stage hindsight.

14.2.1 Perspective

This book is meant to deliver perspective. One can gain perspective through study and doing. Reading this book is study, augmented by our suggested readings (presented at book's end) and/or those discovered yourself. Doing requires experience.

For us, perspective means the deep, broad grasp of a company's instantaneous state, in view of all short- and long-term plans and prospects. The book demonstrates that general company performance is subject to significant design and control, thereby influencing success long before actual results materialize.

But perspective transcends immediate reward and well-laid plans. Success is maximized by, at once, adhering diligently to a plan while occasionally recognizing opportunities that require plan modification.

Wise planners modify plans when apropos. Plan changes, however, cannot be unilateral. You might imagine urgent situations when only one leader is available, but such instances are rare. The rule must be broad top-management review prior to implementing meaningful change.

14.2.2 Planning

Planning is the main theme of Chapter 5. Planning is probably the most valuable time you will ever spend, because proper planning optimizes funding opportunities, objectives, strategies, control, and success. Beware of statements or thoughts that planning is wasteful; this book has shown that such ideas are wrong.

Planning is all the activities used to define, forecast, and anticipate project detail and flow in terms of time and resource requirements. Acquired external facts are combined with internal facts and wisdom to model the future. In short, the known past and the known present, both outside and within the company, are intellectually registered, analyzed, and interpreted to create plans for future action. Prominent reasons for planning are revisited below, now with stabilization and growth stage hindsight:

- *Build the management team.* The management team must be progressively recruited, being complete typically by late concept stage or during the R&D stage. Team excellence and breadth, encompassing all necessary knowledge, skills, and credentials, are essential to meet extraordinary startup growth the company demands. Such demands continue throughout business development, but changing requirements over time can dictate essential changes in management team makeup.

- *Set goals and priorities.* Most basically, planning defines prioritized objectives and strategies for the company and its functions, which must be continuously maintained and reassessed throughout business development. Early planning activities should be molded into a complete strategic planning and management function, probably by the preparation or introduction stage.

- *Compare past and future.* Planning forms a universal realm and language for realistically comparing the past, the present, and the future. This should mature into a main element of strategic management and planning.

- *Forecast and meet human needs.* One facet of planning must gauge future human needs and arrange to meet them. That has become increasingly difficult in modern times, particularly for certain specialties, as product-oriented growth companies face an insufficiently skilled or educated workforce, and there is no obvious relief in sight. In-house training, educational support, improved benefits and workplace enhancements, and other offsetting measures may be obligatory.

- *Forecast and meet resource needs.* Like human needs, future physical resource requirements must be estimated and suitable arrangements planned and implemented. Usually the biggest need is funding, followed by specialized equipment or facilities requirements. Often overlooked avenues are strategic or equity partnerships with suppliers, noncompetitive manufacturers, or special customers.

- *Model future operations.* One visible planning action is building and maintaining current a detailed financial spreadsheet model of present and future business. Multiple model versions to cover various contingencies or interactions with different entities (e.g., potential investors, different professional groups, existing investor reporting, suppliers, strategic partners) may be needed at times. All models must present projections ethically, but appropriate assumptions may vary with the audience.

- *Disseminate information.* Planning must regularly convey information to key company personnel, reported details perhaps varying on a need-to-know or other basis. Accounting must conduct analogous reporting, and coordinated accounting and planning are essential.

- *Measures of performance.* Planning must devise assessments of company progress, necessary for strategic management and planning. Tracking company advancement may require more numerous and detailed indices.

The planning function must do far more than plan. Ensure that planning has sufficient authority and resources consistent with every business development stage.

14.2.3 Management

The mostly temporal book layout did not oblige a separate management chapter. Here we tour diverse topics to emphasize management's critical role.

- *Internal function.* Management must devise, stage, implement, and track all significant functions, large and small. At all business development stages, this embodies C[3] (coordination, command, and control). With company advancement, management assumes increasingly broader roles, as discussed below. Management requires perspective, which, in turn, compels seasoned management. There is only one way to get experience—the entire management team must be recruited fully developed.

- *External function.* The same fully developed management team with relevant C[3] experience and perspective is obligatory here, perhaps more so. A competent, polished, and creative management team is critical for funding program design and conduct and related functions like business plan management, company presentations, and other accompanying elements. Also, favorable professional, supplier, and customer impressions of management often are critical to productive interaction.

- *Company control.* This managerial function transcends internal versus external distinction, to design, effect, track, and adjust plans according to dynamically changing environmental, market, and company dimensions. These factors come under company influence varying from strong to none. With company advancement, we recommend stepwise control enhancements leading to the fourfold control system mix described in Section 13.4.

- *Management makeup.* Team makeup is paramount, of necessity covering every relevant base with top-level credentialed and proven talent. They also must work well as a team, having sincere respect, effective interactions, and compatible personalities in a unified whole perceived as such by employees and knowledgeable outsiders. Plug any temporary gaps with consultants held to the same criteria. The management team must always look and be nearly perfect, and team makeup may need adjustment over time to properly meet existing and foreseeable needs.

- *Strategic considerations.* Progressively mold managerial control to assume full strategic management and planning principles by the preparation or introduction stage. Besides requisite business control, this helps funding, and enhances short- and long-term product development, productivity, and profitability.

- *Management compensation.* The management team could receive salary, equity (ownership) share, benefits, equity options, bonuses, perks, and other elements. Formal employment agreements also can address severance terms, retirement, special insurance, and more. Finally, corporate ownership may be controlled by a buy-sell agreement among key players. Given otherwise equal contributions, less equity goes to progressively later team entrants.

- *Board of directors.* Corporations have a board. One common startup mistake is filling the board with team players; board makeup should feature strong, recognized credentialed individuals typically independent of management who collectively cover major company functions. Build the board like the management team and perhaps let them earn a small equity share, since a board with vested interests is even better. Strongly consider a potential board member's ability to recommend or deliver potential investors. The company's main business attorney may attend board meetings, but board membership might create potential conflict. Ideally, an odd number of board members is best (for tie breaking), three to five members being recommended; one or seven usually is too small or too large, respectively, for efficient action.

Management is a make-or-break function. Do it right the first time; a second chance may not be possible.

14.2.4 Funding

Funding depends on success factor excellence and also must be designed and executed professionally. Short-term funding must be planned and acquired and priced appropriately, fully allowing for later funding needs. Private investor funding is typical at first, involving several distinct funding rounds. Each new round faces fresh challenges, as project activity evolves, along with investor expectations. Eventually, bank financing becomes feasible, which is also very demanding but at least avoids courting individual investors. At every stage, funding fully tests professionalism, success factor prowess, and persistence. The work empowered by every successful funding round primarily must position the company for subsequent funding.

14.2.5 Product

Many chapters in this book focused on a particular business development stage, defined primarily by the state of product development. Here we recast the product and its development, taking advantage of stabilization and growth stage hindsight.

- *User need.* First and foremost, initial and future product must serve defined user functions in a user-desirable way. Key factors include sound value, feature complement, clear and accessible user interface, ease of use, and appropriate functional utility. Also crucial are state-of-the-art quality, safety, efficacy and reliability, complete documentation, regulatory compliance, and environment-friendly components, consumables, packaging, and end-of-life disposition.

- *Market.* Identify tentative target markets early, refining and solidifying the list as product introduction approaches. Early market knowledge is critical to define product features, production design and cost targets, and quality requirements.

- *R&D.* R&D develops the company's first product early in the business development process, fully defining the design approach. Afterward, it is important to keep R&D intact, developing new product versions and exploring and developing new products. One-product companies do not work, because multiple-product diversity is needed to minimize vulnerability and to support growth. Strategic wisdom dictates allocating continuous, meaningful R&D resources, since R&D (and coordinated marketing research) develop all future sales.

- *Design.* Detailed design also must be continuous. For lean operations, the same people may do both R&D and design, shifting emphasis with need. In any case, provision of long-term group resources is important to keep the team together, allowing the company to leverage accrued experience and insight. Consultation can be invaluable for fresh ideas and views or to address unusual requirements.

- *Quality.* Quality should progressively incorporate QC (production policing), QA (company procedure policing), and eventually the dictums of TQM (system of customer-driven company attitudes and practices) as early in business development as logic and resources permit. Ultimately, comprehensive QA operating under companywide TQM should characterize technical product development, design, manufacture, and quality.

- *Manufacturing.* This is the sole source of salable product. The dynamics of purchasing, inventory, production, QC, packaging, distribution, and nonconforming material management increase exponentially with product complexity. Use of modern manufacturing software systems and methods are virtually unavoidable for competitiveness and true TQM benefits. Strategic planning and management should ensure modern manufacturing tools, equipment, practices, and controls from the start.

- *Company contribution.* Treated here is product contributions to company strategic objectives and strategies. It is critical that the product series emanating from R&D, design, and production directly serves identified needs of known customers. Administration's planning employs marketing research, R&D, engineering, and quality inputs and company objectives to specify short- and long-term product detail and direction.

- *Technical support.* More complex products likely dictate a technical support function. This can be dedicated and permanently staffed, appropriate to address steady or high-inquiry volumes. An alternative is to page assigned technical people to handle less frequent inquiries. Assigned people must furnish professional, knowledgeable, complete, and courteous support and problem resolution for all requests.

- *Product support.* A design support function must accompany design engineering. Sold product leads to design issues and problems, requiring assessment and action. Activities include recall, replacement, in-house or field service, design and documentation fixes, and possible legal efforts. Product support is inevitable and requires allocation of appropriate resources.

- *Strategic planning.* The marketing division must participate in company-wide strategic planning. Marketing possesses expertise and experience essential to defining what is salable, how alternatives compare, and what strategies and tactics are needed to effect identified plans.

Products are the company's reason for being. Under TQM, customers define what they want, the marketing division must create awareness and availability, and the technology division conceives, designs, produces, and delivers a quality product.

14.2.6 Marketing

Products may presage company being, but marketing is paramount for realization. Marketing is treated here, looking backward from a stabilization and growth stage perspective.

- *Customer analysis.* This implies thoroughly understanding market(s) of company concern and more. Segment potential buyers into relatively homogeneous groups as an aid to understanding, characterizing each group in detail. As business development advances, successively refine customer analyses to better capture potential buyer profiles, schematic descriptions that are invaluable for ongoing strategic planning and rapid response to unexpected marketplace changes.

- *Marketing research.* Marketing research should be continuous. As stages pass, use accrued data and experience to improve approaches and methods. Also, expand research topics to improve reconnaissance and explore extended product lines or new products and markets for potential future pursuit.

- *Program design.* Marketing must design a program to manage each product, coordinate distinct products in a line, and balance product lines and their components to serve company objectives and strategies. Continuous evaluation of the entire program array then allows tracking progress versus plan, instigating corrective actions as needed, and developing marketing interpretations as input for company strategic planning.

- *Marketing parameters.* Previously described programs include marketing parameters specified and tracked for each product, covering product imaging, positioning, promotion, pricing, and delivery. Parameter sets for each product are designed by analysis and interpretation of marketing research, sales records, customer complaints, comments, and inquiries, in light of marketing expertise. Occasional consultation may prudently aid in-house talent. As stages pass, marketing parameters become better defined and applied, but do not otherwise change character.

- *Design support.* After product introduction, marketing gains real-world product feedback, allowing marketing's support of product conceptualization, R&D, and design to improve. Ongoing marketing research should include added issues raised by the technology division and real-world product use.

- *Product sales.* After introduction, theoretical sales issues become real. Assuming requisite management team marketing talent, that should cause little redirection, but the pre– to post–introduction change is still dramatic. New activities must be managed, including promotion, customer service, order taking and processing, management of delivery (e.g., packaging, distribution, product setup or installation, training), shows and exhibits, program design tracking and correction, and perhaps marketing-related field trails. Marketing must substantially mature as product exposure ensues and grows.

- *Customer service.* This major function arises during the introduction stage. A central switchboard is best, to transfer nontechnical product inquiries to customer service. Key call categories include queries, orders, order tracking, comments or complaints, calls for marketing personnel, and business calls. Resource allocation must cover ongoing inquiry volume at all business development stages.

- *Strategic planning.* Participation by the marketing division in company-wide strategic planning is obligatory. Marketing possesses expertise and experience essential to defining what is salable, how alternatives compare, and the strategies and tactics needed to effect the plan.

- *Division management.* This refers to the marketing division. Factors include general administration, personnel, training, budget management, activity assignments and tracking, and monitoring of division progress against plan.

The marketing division has tremendous influence on company success. Technical people often do not appreciate marketing's professional, highly developed, well-understood field that can orchestrate desired results to an impressive degree.

14.2.7 Location

Location appeared in Chapter 2 as the last of the seven success factors, but it has not been emphasized since. We now detail the issue, from the post-introduction vantage point. Company advancement can accentuate many location factors:

- *Market.* Customer proximity drives some products or industries, but such impact is unusual. Examples include requisite customer and company interactions (e.g., highly complex, custom items), transport issues, extensive centralized training or other support, necessary customer-production interactions, and field trails involving heavy company interplay.

- *Parts and materials.* Bulky quantities or liable production supplies might influence company location, but infrequently.

- *Labor access.* Access to appropriately skilled or trainable labor often is critical. Domestically, this problem has been progressively mounting, is already acute in many areas, and has no obvious relief.

- *Transport.* Materials and goods transport significantly affects many manufacturing and distribution situations. Determine the required transport mode(s) first (e.g., air, road, rail, waterway), then address location relative to necessary production inputs and outputs.

- *Infrastructure.* Some operations may have crucial infrastructure requirements, such as roads, utilities, or special services. Appropriate long-term location might lower production costs or development cost or time.

- *Technical support.* Access to particular expertise can weigh heavily on choices of company or division location. Reasons include proximity to universities for expertise or cheap student labor or access to entities or areas featuring unusual resources.

- *Political climate.* Significant opportunities may accrue from special tax, fee, or environmental incentives offered only in extraordinary locales; distinct laws and regulations may convey similar benefits.

- *Single versus multiple sites.* Multiple company or division sites have appeal for some manufacturing companies. The advantages of a single site include improved economies of scale, reductions in required company infrastructure, more complete utilization of production capacity (switching between products), less intracompany transportation, and likely improved product quality and reliability. The advantages of multiple sites may admit better proximity to markets; compliance with regulations demanding locale content or total manufacturing; improved management of smaller sites; dispersion-reduced vulnerability to political, industrial, economic, or other social contingencies; and company flexibility in exploring alternative operations or working practices and in starting or stopping certain product manufacture.

The impact of location varies widely with company, products, industry, and business development stage. General guidelines are not possible, so management must assess the total situation and design logical objectives and strategies accordingly.

14.3 Where Even the Serious Can Go Wrong

Business development is hard and ruthless but manageable through proper attention and control. Attention is required to fulfill our major points, while control makes such points work in practice. Prominent points were reviewed in Section 14.1, relating how to do things right. This section focuses on defending against primary factor failure.

14.3.1 Company Structure

Inappropriate company organizational structure can be disastrous. New-venture investment is always risky and is worse when investor ownership interests are unclear. Investors usually want an exit mechanism, so freely marketable shares are best. Finally, investors would like brisk, extended growth with eventual public share sales. Many entrepreneurial groups form a partnership or LLC for personal convenience or on (bad) advice. The typical result is extreme funding difficulty and eventual failure. A product-oriented growth company should organize as a corporation to best resolve or avoid the identified problems.

Incorporate and issue stock to the founders long before any funding search, to boost project tangibility and address the following problem. A cash investment immediately revalues all (including founders') outstanding stock to the investor's purchase price, and the capital gain on any stock received at an earlier time and lower price becomes viewed by the IRS as current-year ordinary income. In that way, founders (and any prior investors) face sizable taxes yet receive no money. You might argue that all founder stock covers past and future effort. Alternatively, given a founder-to-first-investor time interval of at least several months, we have been able to defend large founder and investor stock value differences based on significant documented company development between the events. Regardless, with professional legal or accounting assistance, develop a plan that specifically addresses this paramount issue.

14.3.2   Seven Success Factors

Complete all seven success factors perfectly, and success is likely. In turn, you can expend tremendous effort to get five success factors solid, and the remaining two will kill you. Picture an ascending success factor rating scale from zero to 10. A project with all 9s or above is a go. Begin to allow some 8s or worse, and failure risk climbs exponentially.

Business development is quite demanding. That is why we have tagged management team prowess as critical, because weak teams cannot touch 8s or 9s. Even so, given major strength elsewhere, there is one way to offset one or two soft spots: Cover weak zones with a consultant or two. It is difficult to recruit a perfect management team, but all team and no consultants is preferred, so try your best. Then, if necessary, tap the consultant realm to complete the management collective. Consultants may be temporary while you seek to fill management slots, but permanent consultants sometimes are useful. Weaknesses might also be at least partially offset through a strong board of directors or by hiring one or two board members part time for temporary or identified duties.

14.3.3   Planning, the Plan, Funding, and Management

This collection embodies success factor standouts, a common mistake being to downplay the required excellence dictum as follows. It is easy to let planning or the business plan slide as other challenges mount. Funding is usually taken seriously, but people find it hard to believe the formal process outlined in Chapter 7 is absolute. Deviate, though, and your chances in the open investment market essentially vanish. Finally, defying the management team makeup outlined in Chapter 5 is the easiest slight of all. Do so, however, and you fail to get a real team, since it lacks the coordinated knowledge, experience, perspective, interactive skills, competence, and poise needed to orchestrate success. That terminology is certainly correct, because business development success is orchestrated, not won by chance.

14.3.4   Business Development Stages

This book divides business development into stages, to clarify company evolution. Excepting Part II treating obligatory support functions, the book follows the project through company transformation and advancement.

Based on experience, we assigned activities a particular stage of origin (ongoing efforts) or happening (one-time tasks) where occurrence is likely, not academically best. Real-product and business development must contend with temporal, physical, financial, and emotional practicality.

Here lies a potential pitfall. You must be practical, keeping the operation viable, but you also must continually tighten control and otherwise build long-term value. Balance between practical and theoretical, tangible and intangible, and immediate and long-term is one entrepreneurial burden. Not heeding the strategic while tending the current surely erodes strengths and accentuates weaknesses.

A proven way to keep diverse aspects on track is following the plan delineated in this book. Specific functions or actions are tied to identified stages intentionally, because the outlined company development recipe properly orders and organizes activities for competent, efficient realization of interwoven parallel and serial functions. Our scenario is generic, so you may have to adjust the general plan to your unique situation. Otherwise, the basic schedule is highly universal in broad sweep.

Do not depend on luck; stick to conventional wisdom. These points apply to both manufacturing startups and learning chess. When you tackle something new, your initial moves should follow long-proven, straightforward sequences, not adventurous explorations.

14.3.5   Four Control Systems

A significant theme in Part IV has been the importance of four distinct yet interlocking control systems: SOPs, strategic management and planning, QA, and TQM. Human physiology, large jet aircraft, stable ecologies, and other known complex entities employ independent overlapping control systems to establish and maintain coordinated purposeful operation. Your project and prospective company needs the same.

Our recipe starts with one or two simple control means, an early system that is successively augmented over several business development stages into a robust, multifaceted, parallel set of interdigitating control systems. The result is a comprehensive and redundant warning system that permits early identification of problems. Also mandatory is objective and effective problem analysis, corrective action, and followup tracking to ensure problem resolution. These combined interdependent functions provide potent protection against surprise and loss of control.

Businesses become progressively vulnerable to problems, crises, and system failures, to the degree they lack control. Such disasters cost far more than preventative control systems. Control the company, so it will not control you.

14.3.6 Quality

The four control systems embody a global attitude about quality. The uninformed reflexively dismiss quality as an unaffordable luxury, while the reverse represents real truth. Historically, good quality has been shown to save money while improving product perception and utility. Today, quality is not just prudent, it is virtually obligatory.

We have defined four control systems: SOPs, strategic management and planning, QA, and TQM. SOPs and strategic management and planning represent traditional manufacturing company control, written procedures having been popular a very long time, with growing emphasis on formal strategic considerations occurring more in recent decades. The remaining two control systems, QA and TQM, go beyond the traditional. TQM extends quality to all company activities and even core attitudes, by linking thought and action to ultimate customer satisfaction. QA, a method consistent with TQM, is recommended to tighten companywide control of policy, procedures, and behavior.

All four control systems act in concert, system overlap being used for redundant checks and increased efficacy. Overall operations run smoother, cheaper, more productively, and with improved margins. Everyone wins, customers, company, employees, and supplier alike.

We have come full circle. Quality went unmentioned at first, not being mandatory for bare-bones product development. However, it has been discovered that quality-guiding principles for governing decisions and practice makes everything better for everybody. The bystander has become emperor.

14.4 The Entrepreneur's Creed

You have explored high-growth, product-oriented business development in depth; now it is time to explore yourself. Certain qualities are more prevalent in entrepreneurs, especially engineers and other technocrats. Consult the recent book by Edward Roberts, *Entrepreneurs in High Technology*, for more detail, as you rank yourself relative to the following list.

- *Accomplishments.* I rank high against peers at work relative to tangible accomplishments (e.g., projects, patents, publications).
- *Desire.* Starting a project or company has been a long-time personal desire.
- *Experience.* My work experience includes varied roles, settings, and project scope, covering a decade or more.
- *Education.* I have at least a 4-year undergraduate degree or a master of science degree in some technical field.
- *Work ethic.* I am better than average at delivering technical projects and can work hard for something personally important.
- *Work satisfaction.* My current working environment is challenging and brings professional satisfaction.
- *Independence.* My sense of self and independence is strong, and I enjoy successfully overcoming challenges.
- *Advancement.* I have progressed on a moderate to fast career track and have now achieved managerial level.
- *Exposure.* I have had exposure to the self-employed environment, and it seemed less confining and more open to creativity.
- *Financial needs.* My family and I are willing to control expenses or substantially pull back for a sustained period.
- *Demeanor.* I am more outgoing than most engineers but probably subdued compared to extroverts.
- *Orientation.* My background is engineering or another technical discipline, but I think and read more business than technical material.

Individuals moderately to highly consistent with most of the qualities listed are reasonably postured. Key management team members will probably need similar profiles, especially early entrants.

We have taken a comprehensive look at turning a product idea into a growth business, and a wealth of additional information is presented in the accompanying appendixes. Appendix A greatly extends treatment of the business plan. Appendix B features a real-world example of proper business development.

We close with encouragement. Anyone can accomplish business development; ideas can be orchestrated on cue, and everything else subsequently can be taken one step at a time. Surround yourself with top credentialed talent, follow this book's business development recipe, and you will assume a commanding posture for success and its rewards.

Appendix A: The Formal Business Plan

A.1 Introduction

Aligned by management perspective, the planning function of the administration division sets the philosophy, course, and rationale for all activity. Applying strategic planning principles, management develops 5-year and annual objectives, the strategies to carry them out, and specific tactics for implementation. The business plan (the plan) is a formal written document that summarizes a selected subset of historical and planned activities for a stated time and audience.

Chapter 6 focused entirely on the business plan, including its nature, rationale, content overview, and physical assembly. Section 6.8 emphasized that the plan is really many plans, because numerous updates usually arise during extended business development. In addition, Section 5.3 outlined the general format and makeup of spreadsheet software typically used to build the financial model.

This appendix greatly extends Chapter 6, defining business plan organization and content down to individual topic levels. The unique business plan format described in Chapter 6 is assumed. Addressed in turn are the three prime plan divisions, the narrative, financial model, and appendixes.

Business plans serve a multitude of purposes. There are four primary reader tests a business plan must pass:

- *Five-minute test*: a rapid overview to permit multiple plans to be sorted, tested against defined criteria for applicability, or as orientation for deeper perusal;

- *Fifteen-minute test:* a more thorough overview to retest applicability, assign detailed review responsibility, or as broader orientation for detailed perusal;
- *Selected in-depth topics:* typically used to grade the project against business development criteria to assess project merit;
- *Full reading and study:* comprehensive study and evaluation of the entire plan.

Most people new to business development assume the business plan will be fully read. However, complete readings are infrequent, usually because of a poor plan or project. Well-designed projects with well-written plans command significant attention.

A.2   Business Plan Narrative

Table A.1 presents the business plan table of contents for a nominal product-oriented, privately funded, high-growth company early in project tenure. The plan narrative includes the Executive Summary and seven sections: (A) Program Overview, (B) Management and Strategies, (C) Products, (D) Marketing and Sales, (E) Competition, (F) Operations, and (G) Risks and Rewards.

A.2.1   Executive Summary

The executive summary is designed to pass the five-minute business plan test. This section features eight topics, typically involving one paragraph each.

- *Overview.* Name the organization and generically describe its nature, mission, long-term objectives, and products.
- *Competitive products.* Describe product types available from competitors and their weaknesses compared to your products, as lead-in to the next topic.
- *Your products.* Depict your existing and future products and how they improve competitive shortcomings (list only the best one to three products here; save the rest for the next topic).
- *Added product issues (optional).* Go on to describe additional benefits of your products and also honestly mention any weaknesses and how they are being addressed.
- *Business development status.* Briefly outline the status of both product and business development, the latter one individually for each of the three divisions (administration, technology, and marketing).

Table A.1
Table of Contents for an Early-Stage, Product-Oriented Business Plan*

TABLE OF CONTENTS	
Executive Summary **A. Program Overview** Introduction History Products Market Marketing and sales Competition Location Funding requirements Financial goals Return to investors **B. Management and Strategies** Corporate mission Business development strategy Goals Management team Consultants **C. Products** Concept Product approach Product design Patent, trademark, and regulatory status Future products	**D. Marketing and Sales** Marketing plan overview Sales plan overview Market size Marketing research Marketing for the initial product Future markets and their approaches **E. Competition** Perspective Specific competitive products Anticipated competition **F. Operations** Perspective Product R&D Product design Manufacturing Quality **G. Risks and Rewards** Risks Benefits Stockholders

* Appendixes are not listed here due to their inherent variability (see Section A.4).

- *Market(s) and shares.* Profile the market(s) (the nature and size) your products address, extracting summary data from the program overview and the marketing and sales section.

- *Project strengths.* List the major strengths of the project (e.g., management team, product features and acceptance, patent status, technical prowess, marketing prowess).

- *Funding requirements.* Describe the funding sought, how it is to be used, how it relates to total long-range funding plans, and what is offered in return. In a brief table below the ending text, summarize for each year of the 5-year model: (1) the required investment, (2) sales, (3) market share, and (4) pretax profit.

The length of the executive summary should not exceed one and a half pages.

A.2.2   Program Overview Section

The program overview supplies a crucial transition between the brief executive summary and the remaining six sections of the narrative. It covers the topics that follow (following each topic is the suggested length):

- *Introduction.* The introduction provides an orientation to the broad industry or industry group, describes the more specific industry of the company product, depicts the base technology's state of the art and its strengths and weaknesses, and closes by reviewing the product and how it exploits or extends the technology. (4 to 5 paragraphs)

- *History.* Include here a nondetailed list of people recruited and goals achieved to date, such as pertinent history, technical and marketing development, management team status, facilities, patent status, and project recognition. (1 to 2 paragraphs)

- *Products.* Describe what the product accomplishes functionally (not what it is), summarize the product's technical rationale, and list product attributes labeled important by industry representatives. (3 paragraphs)

- *Market.* Depict the market size, backed (if possible) by references, market share estimates, future expansion of markets with the same or other products, and why the market is considered appropriate for the product. (3 paragraphs)

- *Marketing and sales.* Very briefly, summarize marketing objectives, strategies and their rationale, near-term promotion and sales activities and their assessment measures, product imaging and positioning details and associated rationale, and the expected long-term sales growth profile and its rationale. (4 paragraphs)

- *Competition.* This topic is a substantially condensed version of the full section on competition. Modify the presentation format to reduce unavoidable redundancy. (1 to 2 paragraphs)

- *Location.* The primary discussion of facility location is here, so cover future plans as well as past and current data. (1 paragraph)

- *Funding requirements.* The main development of funding requirements resides here. An overview paragraph introduces a table that shows, by project quarter, funding needs and a condensed summary of principle company activities. Parenthetically, irregularly breaking up the narrative with a table, a figure, or a list aids readability.

- *Financial goals.* After an introductory paragraph citing the financial model as source and briefly listing key model assumptions, provide a tabular summary of the following key financial goals for each year of the plan: revenue, COGS, expenses (all, lumped), pretax profit, taxes (at an assumed rate; consult your accountant), after-tax profit, number of outstanding stock shares, and earnings per share.

- *Return to investors.* Summarize the general distribution of stock among investors and management, on the basis of the data in the appendix on organization (see Section A.4.1). List and briefly discuss (typically) three mechanisms identified by management as exit strategies feasible several years out (e.g., company stock buyback, private offering, public offering, company borrowing, merger, company sale).

Design the program overview section so that discussion does not duplicate discussion elsewhere. Many general topics are specified repeatedly in the plan, but each occurrence calls for a different approach or distinct level of detail. No direct rehashes of previously presented material should occur anywhere in the plan (except perhaps the executive summary).

A.2.3 Management and Strategies Section

Most business plans slight the management team. That mistake overlooks a typical investor's prime management team interest, which often exceeds the product itself. We place the management team first after the program overview, adding a related discussion of strategic concerns to further strengthen the presentation.

- *Corporate mission.* In one or two sentences, state the corporate mission exactly as needed for ongoing strategic planning (see Section 11.4). (1 paragraph)

- *Business development strategy.* Outline company and product development and underlying rationale, describe industry reception or pre-introduction marketing research to date, discuss how you have applied the strategy of customer-driven design approach, and outline future product development plans and their dependence on the same principles. (4 paragraphs)

- *Goals.* State the model's fifth-year sales goals, followed by a bulleted list identifying at least three reasons that should materialize (e.g., your unique technology, extensive team industry knowledge, current market position, new or extended marketing approaches planned, future products and markets). (1 paragraph)

- *Management team.* List the current team and relevant background or an overview to lead into the following bulleted list (1 paragraph). List each team member, followed by a brief summary of his or her professional field, specialty, major strengths, education, and role (1 paragraph per member). Finally, include a closing paragraph summarizing the time commitment and strength of project conviction of all players, individually or collectively. (1 paragraph)

- *Consultants.* If you agree, add something akin to this: "[*name of company*] ascribes to the concept of using consultants instead of employees for certain tasks. This reduces long-term commitments and therefore costs for irregular or low-level activities and, if relevant, allows evaluation for potential hire. Funds for many in-house administrative, technical, and marketing/sales tasks may be handled using consultants rather than staffing." (1 paragraph)

A.2.4 Products Section

Most product discussion resides here, with emphasis on current and near-term activities. The products section is presented early to accommodate high investor interest.

- *Concept.* Describe the fundamental concepts that underlie early products, features that embody said concepts, and expected customer reactions. Outline any significant future enhancements or additional concepts. (1 paragraph)

- *Product approach.* Review the specific design approach selected for the first product, distinguishing this discussion from that immediately preceding or following. (1 paragraph)

- *Product design.* Describe the initial product design in some detail (without revealing proprietary elements) as follows: general nature of product, design status, intended use, configuration of various models, a bulleted list of major industry design and functional improvements accomplished (e.g., simple mechanisms, easily produced, reliable components, extended guarantee, and so on), and means of production. (5 to 6 paragraphs)

- *Patent, trademark, and regulatory status.* Review patent status of near-term products, singly or collectively as appropriate; in turn, do the same for trademarks, service marks, and regulatory issues. (1 to 3 paragraphs)

- *Future products.* Thoroughly describe the philosophy, organization, budget weight, and leadership of R&D and present its long-term project plans.

A.2.5   Marketing and Sales Section

The marketing and sales section contains the most discussion of marketing and sales, with primary attention to current and near-term activities.

- *Marketing plan overview.* Begin by stating your ultimate realistic industry goal (e.g., leader, prominent or general participant, number two player). Name the functions under the control of the marketing division (probably program design; customer analysis; product imaging, positioning, promotion, pricing, and delivery; competitive analysis; forecasting; new product planning; sales management; and customer service) and give a brief description of each in bulleted list form (1 paragraph plus the list). Describe near-term activity focus and lead into the following four broad discussion topics (1 paragraph). Address the marketing portion of corporate strategic management, directed to the strategic and tactical planning to develop, design, implement, and control all marketing and sales functions, and who will control that function (1 paragraph). Treat product marketing, the function that defines system features, customers and their needs and desires, product imaging and pricing, sales leads, forecasts, product and technical literature, and other promotional tools and who will control that function (1 paragraph). Next, address promotion and advertising by describing current or planned elements, the target marketplaces, and how your marketing properly conforms to said markets (1 paragraph). Finally, mention sales and service purely by defining management names and titles. (1 paragraph)

- *Sales plan overview.* Review the general nature, structure, leadership, makeup, and compensation structure of the sales function (1 paragraph); outline the general approach and steps to consummate typical sales (1 paragraph); and describe customer service function leadership, structure, size, makeup, methods, and how data are collected to aid product design. (1 paragraph)

- *Market size.* Define market size and relevant assumptions; present a table of market size, company sales, and company market share for each plan year; discuss reasonableness of such company market share achievement; and end by listing three strong attributes of your product fostering market penetration (e.g., value, low cost, improved efficiency). (4 paragraphs)

- *Marketing research.* Subtly review company and product advantages from a different perspective. Introduce and display a bulleted list of primary concerns expressed by industry reps (or customers), then present the list, choosing items that accentuate your positives as learned from marketing research (e.g., customer needs, desires, concerns, and short-comings; appropriateness of industry sophistication, diversity, and maturity; environmental or regulatory issues where you shine), and finally close with a brief description of how well your product meets list attributes (so list only items where you excel). (2 paragraphs plus list)

- *Future markets and their approaches.* Briefly describe several future product and/or marketing approach concepts. Keep this subsection short, to avoid imbalance and a "dreamer" image, but it is critical to show your essential commitment to ongoing product development. (1 paragraph)

A.2.6   Competition Section

Information of the primary competition resides here. Competitive data get slighted in many business plans, but investors consider the subject critical.

- *Perspective.* Provide an overview of the global market (size, growth trends, nature of customer, product turnover rate and so forth). Then describe how the material to follow is organized, for example, by product within one market (as is assumed in the next item), by niche within one market, or by markets.

- *Specific competitive products.* Whatever organizational approach is chosen, discuss each competitor by name, addressing size, company age, spectrum of products, competitive strengths and weaknesses (e.g., product development, innovation, design, quality, marketing), products specifically competitive to yours, and the good and bad of theirs versus yours. (1 paragraph per competitor)

- *Anticipated competition.* Describe company preparedness for competition in general, including future new market entrants. It is best to emphasize (and actually conduct) continuous marketing research, in-house R&D, aggressive product development and market entry, and any other factors relevant to sound, responsive business and product management.

A.2.7   Operations Section

Primary information about internal technical operations is placed here. Many business plans omit or slight this discussion, but your approach and methods of design and production are essential to complete project evaluation.

- *Perspective.* Introduce and define in logical order each major company technical function for later discussion. We treat functions in order of product development, with quality last to emphasize its central attention. (1 paragraph)

- *Product R&D.* Describe product development organization, its leader, to whom the leader reports, facilities and equipment, functions included here, sources of ideas, and other key data concisely conveying how R&D works and meshes with other company functions. (1 to 2 paragraphs)

- *Product design.* Review the same basic topics as R&D, plus how a design moves through the system and documentation and change control management. (1 to 2 paragraphs)

- *Manufacturing.* Treat the same basic topics as R&D, plus how key elements (parts, materials, progressing product, finished product, nonconforming material and so forth) flow through the system, inventory management, and packaging and shipping. (1 to 2 paragraphs)

- *Quality.* Discuss the same basic topics as R&D, emphasizing how quality is a distinct function and how it is kept relatively independent of management of the administration division and the technology division (1 paragraph). The latter treatment concerns quality at the departmental level. Then, assuming adherence to this book's recipe, treat each of the two major quality functions, QC and QA, in some detail, describing their organization, leadership, functions, facilities and equipment, and other key issues.

A.2.8   Risks and Rewards Section

If you are seeking private funding, this section moves from being important to being obligatory. Assuming no exposure of the business plan to potential investors, the subsections on benefits and stockholders could be eliminated. That is not recommended, though, because unexpected exposure or altered plans regarding funding may arise.

- *Risks.* Begin by asserting that it is essential for management to identify and assess both the risks and the rewards of the company's venture, to aid informed planning. Then state that the plan does not constitute a solicitation or offer of investment—the presented treatment is for purposes of planning and information only, and detailed investment data will be provided during relevant deliberations. Include the same statement on the frontmost cover or title page of the business plan as well (1 paragraph). Next, include a bulleted list, with each item formed as follows: Identify one major project risk, followed by a brief explanation of how the project is addressing that risk (1 paragraph for each list item). Risk examples include management's ability to effect successful administration, marketing, sales, and products; the degree to which company products will be accepted by customers and bring them benefits; management's ability to provide adequate external funding and proper financial management; unexpected social, economic, regulatory, or legal perturbations; the advent of competitive responses, products, and firms; and copying of company products or technology.

- *Benefits.* Briefly discuss or list potential benefits to founders (e.g., salary and benefits, entrepreneurial satisfaction, security, stockholder benefits), employees (e.g., salary and benefits, job satisfaction, advancement and professional development opportunities), and stockholders (e.g., potential equity growth, potential dividends, entrepreneurial satisfaction).

- *Stockholders.* To elaborate stockholder risks versus rewards, investment in startups or young firms usually accesses potentially rapid increases in net worth without an immediate heavy tax burden. While Congress has removed some advantages, it is still possible to substantially build investment net worth, without tax payment until ownership equity sale. You can even predict company book value (company valuation divided by number of shares outstanding) and market value (company fifth-year after-tax profit multiplied by an assumed price-to-earnings ratio). For the latter calculation, the price-to-earnings ratio typically is selected in the range of 5 (mundane products or industries) to 10 (exciting, leading-edge products or industries). If you display projected market value in this way, also state the assumed price-to-earnings ratio.

Writing a strong business plan is difficult for most people; it calls for special knowledge uncommon to education or the workplace. Generation of a business plan requires skilled technical writing, broad business knowledge, and knowledge of audience wants and needs. The generic narrative described here and the generic model and appendix descriptions to follow provide a solid basis for writing any product or service business plan for any stage of company development. Only extraordinary circumstances should dictate much need for adaptation.

A.3 Financial Model of the Business Plan

In our method, the completed financial model occupies the first two appendixes, being considered too detailed and bulky for the business plan narrative. We discuss bare model essentials here, the financial assumptions and the spreadsheet model proper. Other accompanying elements are introduced in Section A.4.

A.3.1   Model Assumptions

It is paramount to present model assumptions prior to the spreadsheet model proper. Spreadsheets are meaningless without specifying the assumptions involved in their creation. Nevertheless, few plans discuss relevant assumptions. Listed assumptions should address all financial statements presented, not just the primary (income statement) schedule. Topics for a bulleted assumption list might include:

- Key assumptions used to develop the sales projections;
- Marketplace and competitive assumptions;
- Pricing and discount policies and rationale;
- Rationale for salaries, benefit levels, and other compensation;
- Average time lag between a sale and actual money receipt;
- Average time lag between incurring and paying of bills;
- Major bills or expense categories that cannot be postponed;
- Functions that are contracted rather than conducted in-house;
- Rationale for leasing or purchasing large-ticket items;
- Time-dependent relationships, such as volume purchases;
- Relevant environmental, social, economic, political, and regulatory issues.

Despite the length of the list, this topic should be short. One page is best; beyond two pages is excessive.

A.3.2   Spreadsheet Model

This discussion employs the following specific structure for illustrative financial statements:

- Headings are in bold type.
- Line items under each heading are presented in lower case.
- Totals and subtotals are underlined.

The following subsections examine generic versions of the three primary financial statements—the income statement, the balance sheet, and the cash flow analysis—in some detail.

A.3.3 Income Statement

The income, or profit-and-loss (P&L), statement summarizes income versus expenses. To hold this statement to one page, many data are summarized, which often dictates the need for additional statements extending beyond the three primary ones (see Section A.4) to present supporting detail. Otherwise, readers of the business plan may erroneously judge primary statement numbers as too arbitrary.

Table A.2 is a layout for a generic income statement for a product-oriented growth company after product introduction (thus, sales activities are included). Terminology is consistent with the discussion in Chapter 5 of the spreadsheet model. The table indicates row headings only, the spreadsheet structure being completed by defining column headings, typically one column for each quarter of the 5-year plan, plus an annual total for each year. For programming ease, place all annual totals in adjacent columns at the spreadsheet's far right, rather than respectively positioning an annual total after the four quarters for each year.

A.3.4 Balance Sheet

Presented immediately after the income statement, the balance sheet shows the relative balance among assets, liabilities, and company equity. A balance sheet is organized to equate assets with the sum of liabilities and equity, the latter term being the company's net worth. Thus, the balance sheet summarizes the company's general financial health, its history, and its projected future evolution.

Table A.3 is a layout of the row headings and line items of a generic balance sheet. It is consistent with the income statement in Table A.2 and the spreadsheet terminology in Chapter 5.

A.3.5 Cash Flow Statement

Our recommended primary financial statements also include a cash flow statement presented immediately after the balance sheet. There are several cash flow statement formats, but for operations using cash accounting, all formats simply summarize the actual flow of cash in and out of the company.

In the latter respect, cash flow statements highly resemble the income statement data, which also display the actual flow of cash in and out of the company. Nevertheless, a small number of items that may appear on cash accounting income statements do not represent immediate flow of cash, so a separate cash flow statement is used to provide a view restricted to cash flows. The following is a nonexhaustive list of some such income statement items that do not relate to immediate movement of cash:

Table A.2
Income Statement for a Generic Product-Oriented Growth Company

Revenue	
Net sales	Total of all money received
Other revenue	A net sales breakdown by component is possible too
Total revenue	Total of all money to the company
Cost of Goods Sold	
Contract manufacturing	Whole product or subassemblies
Purchased materials	All purchased parts and materials
Manufactured materials	All parts or materials made in-house
Labor	All direct/support labor involved in manufacturing
Total COGS	Total of all COGS
Gross margin	Total revenue less total COGS
Direct Selling Costs	
Sales expenses	All direct expenses of selling
Commissions	Sales commissions and bonuses
Ads/meetings/shows	Self-explanatory; add rows for finer breakdown
Selling costs	Total of all direct selling costs
General and Administrative	
Administration labor	Labor + personnel overhead for admin. division
Administration expense	All nonlabor expenses for administration division
Technology labor	Labor + personnel overhead for technology division
Technology expense	All nonlabor expenses for technology division
Marketing/sales labor	Labor + personnel overhead for marketing division
Marketing/sales expense	All nonlabor expenses for marketing division
Rent, utilities, phone	Include equipment under appropriate division
Legal, insurance, accounting	Self-explanatory
Other	Break into additional line items as needed for clarity
Total G&A	Total of all G&A expenses
Profit & Loss	
Profit before tax	Algebraic sum of all category headings
Less tax @ 37%	Consult an accounting professional for the correct percentage
Profit after tax	The infamous "bottom line"

- Depreciation expense (an allowance, not an actual cost);
- Estimated taxes due later;
- Appreciation of assets (again, an allowance, not a cost);
- Interim accumulations to offset periodic payments of personnel benefits;
- Payments into a self-insurance fund.

Table A.3
Balance Sheet for a Generic Product-Oriented Growth Company

Current Assets

Cash	Cash, typically the checking account balance
Accounts receivable	Money due the company in normal business conduct
Raw materials inventory	Inventory as parts and materials
Work-in-progress inventory	Inventory of partially assembled goods
Finished product inventory	Inventory of finished product fully ready for sale
Short-term notes receivable	Short-term notes payable to the company
Total current assets	Self-explanatory

Long-Term and Fixed Assets

Technological rights	Noncash value conveyed by founders, for stock
Equipment	Purchased large-ticket items
Less depreciation	Depreciation of equipment (a negative entry)
Improvements	Improvements made to facilities
Long-term notes receivable	Long-term notes payable to the company
Total L/T&F Assets	Self-explanatory
Total assets	Total of all assets; equals liabilities + equity

Current and Long-Term Liabilities

Accounts payable	Money owed by company in normal commerce
Short-term loans	Short-term loans received by the company
Tax, payroll	Payroll taxes
Tax, other	Other taxes; break down if needed for clarity
Total current liabilities	Total current liabilities
Long-term liabilities	
Total L-T liabilities	Long-term liabilities owed by the company
Total liabilities	Total current + total long-term liabilities

Equity

Common stock	Founders stock + stock in return for investment
Preferred stock	Often this is zero for all time periods
Retained earnings	Cumulated excess earnings, per income statement
Total equity	Algebraic sum of all equity
Total liability and equity	Algebraic sum of all liabilities and equity

Note that some, but not all, of those items appear in the sample income statement (Table A.2).

A.4 Business Plan Appendixes

Appendixes allow inclusion of selected details in the business plan without burdening the narrative. Readers can consult just those appendixes of interest if and when needed.

Every business plan should include the following core appendixes (Appendix III only if it is relevant):

- *Appendix I.* Primary financial projections (income statement, balance sheet, cash flow analysis);
- *Appendix II.* Supportive financial projections used to elaborate summarized material in the primary financial projections;
- *Appendix III.* Historical financial data (optional; included only if relevant);
- *Appendix IV.* Company organization; a place for accessible extra detail regarding company internal structure, incumbent leaders, board members, and active and planned stock distribution;
- *Appendix V.* Resumes of the management team and the board of directors.

The following material first addresses the five core appendixes and then examines several appendix categories contingently extending beyond the core group.

A.4.1   Core Appendixes

The five (or four, if Appendix III is omitted) core appendixes should be placed first among all appendixes, in the following order:

A.4.1.1   Appendix I: Primary Financial Projections

The first appendix should always include the income statement, the balance sheet, and the cash flow analysis, each filling one page in landscape (broadside) presentation mode. Prior to those three figures, begin the appendix with a narrative introduction and the model assumptions. The introduction might review company financial strategy, accounting type (cash or accrual), how time periods are set, how funding will be employed (presumably according to the model), and a description of appendix contents. Alternative model time interval formats include specific dates (e.g., "1st quarter 2002") or contingency on events (e.g., "receipt of funding"). Examples of the latter include quarters triggered by first funding, a prespecified funding threshold, or receipt of all current-round funding. For contingent timing, label columns as 1st Quarter, 2nd Quarter, . . . Nth Quarter or Q1, Q2, . . . Qn.

A.4.1.2   Appendix II: Supportive Financial Projections

Technically, this appendix is optional but strongly urged. We universally use and highly recommend the following three distinct supportive financial statements, all covering the full 5-year plan and employing the same column heading layout as the primary financial projections (typically quarters, with all five annual totals grouped at the far right).

- *Sales forecast.* Expand the sales presentation to a full landscape-mode page layout, which gives plenty of room to break down overall sales into many informative subcategories. Also included for increased realism might be independent rows (line items) to address warranty and shipping costs, bad debt allowance, and other revenue adjustments.

- *Nonpersonnel expenses.* Placed here are all expenses not directed to fixed salaries or hourly wages. One good layout is by company division, department, or other functional breakdown. All consultation should be listed here, associated with the applicable functional group.

- *Staffing forecast.* This financial statement can provide an excellent view of staffing plans. The best organization is to duplicate the functional frame chosen previously for the nonpersonnel expenses. This statement should include a separate breakout for personnel overhead, so the reader can specifically see how benefits and other personnel costs affect operations.

A.4.1.3 Appendix III: Historical Financial Data

For projects with a financial history, present relevant summarized data to demonstrate the flow of money to and from the company. This appendix might be presented as distinct statements for each year, as a time-based table like the other core appendixes, or some other informative format. The format here usually need not match other statements.

A.4.1.4 Appendix IV: Organization

The management and strategies narrative section briefly addresses certain organizational issues, but it is brief to preserve narrative balance. The major discussion occupies our sample Appendix IV (Appendix III, if the optional historical appendix is omitted). Following a brief introduction, provide a bulleted list designating and describing each of the following elements (several lines per heading).

- *Board of directors.* Size, functions conducted, advisory only or active on certain issues, incumbents (place resumes in Appendix V);

- *President.* Incumbent, typical responsibilities, to whom/what president reports (the board, presumably), general philosophy of management and/or business development, and/or other key elements;

- *Other management.* Basically the same outline employed for the president, separately for each key management team member;

- *Consultants.* Consultants used to cover essential management gaps or to address major extended issues analogous to management.

This appendix provides considerable increased understanding, with little overlap to the narrative's management and strategies section. It should outline the company's general ownership structure (who owns how much, allowances for future funding and other stock purposes), the nature of any stock earn-out arrangements, and any active buy-sell agreements or other ownership constraints.

A.4.1.5 Appendix V: Resumes

This appendix is critical to document management and board credentials. Potential investors rarely meet management team members prior to reading the business plan. Thus, impressive face-to-face demeanor, knowledge, and experience are not enough. Every team member also needs strong credentials, and the resume must clearly reflect credentials, knowledge, and experience in impressive fashion. Insist on strong resumes and rebuild any that do not exude the appropriate strength.

A.4.2 General Appendixes

These business plan supporting elements vary much more in occurrence and makeup than do the core appendixes. The discussion to follow focuses on appendix categories, since individual details are too variable for rigorous specification.

- *Technical.* Design documentation, product test data, design verification and validation data, product cost estimates, production procedures or plans, quality procedures or data, competitive product technical analyses, technical support plan, technical specifications;
- *Marketing.* Marketing research results, detailed marketing plan, customer characterization and segmentation analysis, competitive product marketing analysis, promotional material copies, product imaging and positioning analysis, pricing plan, distribution plan, marketing specifications;
- *Sales.* Detailed sales plan, sales regional structure, sales methods specification, forecasting methodology, customer service plan;
- *Administration.* SOPs, strategic planning and management function, TQM plan, human resources plan, management issues, corporate formation documents;

- *Quality.* Quality plan, quality control plan and procedures, QA plan, competitive product quality analysis, quality specifications, nonconforming material procedures;

- *Literature.* Key articles from the literature, technical bibliography, technical standards;

- *Photographs.* Prototypes, finished company product, competitive products.

There is no magic number of appendixes, but too many detract from the business plan. Make sure that the appendixes collectively cover all requisite business plan support issues and that every appendix is important. The exact appendix count is irrelevant.

Appendix B: War Story: All the Right Stuff

B.1 Introduction

This section is about a winner, a project in which everything was done right, based on proper performance and advancement. The company is ThermoGear, Inc. (ThermoGear™), 18005 Lower Boones Ferry Rd., Tigard, OR 97224 (503-624-1415; chillbuster@thermogear.com).

The inaugural ThermoGear™ product is a battery-powered, rechargeable, portable electric blanket system (Figure B.1), to be promoted under the ChillBuster® trademark. The system is composed of five elements:

- *Blanket.* The ChillBuster® product is technically distinct from but functionally analogous to a standard electric blanket.

- *Control module.* The control module comprises an electronic circuit board with all operating and control electronics, a user interface panel, appropriate interconnect wiring, an enclosure for the latter elements, and an external cable extending to a blanket-proximate connector.

- *Battery.* The control module holds the separate system power source, a rechargeable battery.

- *Cabling.* An optional cable connects the control module to a vehicle cigarette lighter socket, for external system operating power or system battery recharging.

- *Carrying bag.* Included is a custom tote bag that organizes system components and permits easy transport.

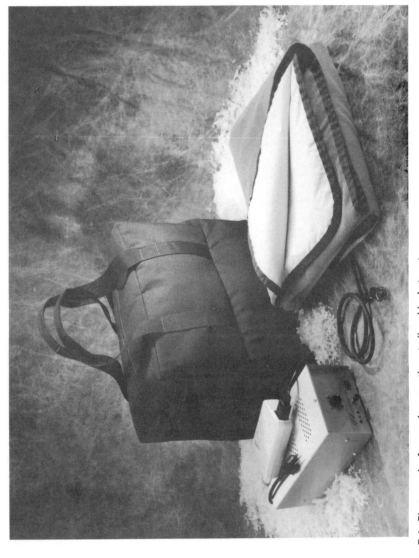

Figure B.1 Photograph of second-generation stadium blanket system.

Special system features form the basis of pending U.S. and foreign patent actions.

ThermoGear™ plans to build an entire active-heat (powered) ChillBuster® product line, accompanying other non-ChillBuster® products, over time. Multiple products and lines are important, because single-product companies typically fair poorly in business and funding.

ThermoGear™ was formed as an Oregon corporation in January of 1997, the project having advanced to the point that it required outside funding. Incorporation was deemed critical to best facilitate the funding process. A C-type rather than an S-type corporation was elected, because one potential early investor was located in Canada (S-type corporations cannot have foreign stockholders).

Incidentally, C-type corporation status is automatic upon incorporation. You must apply to the Internal Revenue Service to request S-type status, and the filing must meet certain time windows and other criteria. ThermoGear™ simply completed the Oregon state procedures for incorporation, applied for a U.S. federal Establishment Identification Number, or EIN, and began regular IRS tax filings as a C-type corporation. ThermoGear™ had no employees, so beginning normal employee-dependent reporting and payments was not required.

Here we provide an overview of ThermoGear™, with further details supplied throughout this appendix. The company originally planned to manufacture and market products conceived and developed in-house. In fact, an initial portable electric blanket product underwent limited sales in a 1997 Christmas catalog. However, two opportunities arose that changed that original thought. At least near term, the ThermoGear™ vision centers on two primary objectives:

- *Nature of the company.* Management found a New York marketing company (anonymous, until relations are consummated) whose product lines and sales channels were well matched to ChillBuster® blanket products and other ThermoGear™ ideas. The two groups are exploring a long-term, mutually beneficial relationship. Arrangements with other groups also are being explored. ThermoGear™ now sees itself as an R&D company, at least for awhile, developing licensable products for manufacture and marketing by others. The original plan to become an in-house manufacturing and/or marketing company may be revisited later. Management remains flexible to any viable approach.

- *Nature of the product.* ThermoGear™ had always intended to drastically upgrade the early test-marketed product from fabricated aluminum to an injection-molded control module as soon as resources permitted. However, based on insightful marketing research, the New York marketer discovered challenging retail price targets dictating immediate need for molded product. Thus, ThermoGear™ started from scratch to design a tightly cost-controlled, molded control module product. The redesign was completed in early 1999.

The New York entity alignment precipitated extensive redesign and delay but also conveyed two major strategic advantages. First was invaluable design specs from direct major retailer feedback inaccessible to ThermoGear™ on its own. Second, once ThermoGear™ becomes ready, it will enjoy instant access to several large distinct marketing channels.

Strategic alteration of ThermoGear™ plans was pivotal. Normally, plans should be a sacred, steadfast guide for thought and action. Occasionally, though, plan changes can be beneficial.

B.2 Project History

ThermoGear™ began with a classic primordial vision, the primal flash. Alumnus Mel Campf was attending a late 1995 University of Oregon football game. It was a cold, rainy, and windy day, very uncomfortable for Mel. With wandering thoughts between plays, he imagined a better means of warmth besides heavy clothing. A portable electric blanket was only one of many alternatives, but this particular idea resonated in his mind, persisting for days.

B.2.1 Bench Prototype

Mel's initial investigation revealed the apparently obvious reason that no portable electric blankets were available. The power requirements are high, compelling too large (and heavy) a battery. That did not deter Mel, who did not know the limitations for sure. He began detailing system requirements for a workable stadium blanket.

By the fall of 1997, Mel had done much research and completed one development cycle. The bench prototype was a proof-of-concept demonstration lacking commercial potential, but it advanced the project nonetheless. System feasibility seemed possible, if certain power-saving means were employed.

B.2.2 Design Company Prototype

Mel correctly set aside a moderate stock fraction to recruit a strong management and product development team. While searching for appropriate team members, Mel retained a professional design group to produce a second-generation design and prototype. During the design effort, Mel encountered this book's author,

Wayne Fields, who instantly recognized significant product potential and also Mel's business strength, two essential success factors. Wayne joined the team strictly for a project equity share, knowing that cash is the most precious startup resource.

Mel and Wayne continued to oversee the design group, and they ultimately obtained one prototype and useful but incomplete design documentation. Mel, with Wayne's help, mounted detailed searches to locate various system component contract manufacturers. In parallel, Wayne developed a formal business plan for a company appropriate to shepherd this and future products. Shortly after completion of the business plan, Tom Lewis, strong in business, entrepreneurship, and financial management, joined the project team.

Four major factors followed in step. First, Tom was not only critical professionally, he also possessed potential investor contacts. Those investors proved interested and worthy, and they placed sizable investments in ThermoGear™ in early 1998 and more later. Mel and Wayne found additional investors over time, and the project has managed to stay abreast or ahead of needed funding ever since. Second, Mel and Wayne had developed additional product ideas for ThermoGear™, and the diversified potential product mix conveyed significant influence in attracting further management team members. The third event resulted from contract manufacturing searches. Management uncovered Jack Robinson, a talented entrepreneur and manufacturing company co-owner skilled in fabric-based products who joined the ThermoGear™ team. Fourth, through referrals, management located Marc Ranger, a talented senior electronics engineer who initially was to help bring second-generation system electronics and prototype design to production status. Marc has provided stunning contributions to design evolution and revolution ever since, and he was the last of the current ThermoGear™ management team to come aboard.

B.2.3 Production Design

The ThermoGear™ engineer, Marc Ranger, first worked as a project consultant during the design company prototype (second-generation design) phase. Shortly thereafter, management encountered a major New York marketing concern with mutual interests. The New York group had several established marketing channels pertinent to expanded ThermoGear™ plans. In turn, the New York group was very interested in the stadium blanket and other ThermoGear™ product ideas. The New York group used the second-generation stadium blanket design to explore product interest with major domestic retailers. The result was the aggressive pursuit of a third-generation design just ready (fall, 1999) for testing, salespeople's samples, and eventual volume production.

ChillBuster® blanket design now stands as follows. The control module, cabling, ac adapter, and carrying bag have been designed, and contract manufacturers have been secured. Blanket design proved to be a major problem, seemingly due to the public scare about electromagnetic radiation a few years ago. That resulted in closure of most domestic electric blanket production plants, causing a dearth of potential ChillBuster® blanket contract manufacturers. After much difficulty, a primary blanket contract manufacturing arrangement was identified and prototypes verified. Management has also developed two backup possibilities, should the primary arrangement prove not acceptable.

B.3 Management

Management is one of the seven success factors introduced in Chapter 2. Summarized below are the ThermoGear™ management team, their roles, and the particular strengths each brings to this exemplary entrepreneurial project.

- *Mel Campf, president.* With normal duties accompanying such office, Mel is also chief author on all patent actions to date. Mel's strengths are in corporate management and program design and implementation, with particular expertise in marketing and sales. Mel also is co-owner of a separate retail business.

- *Wayne Fields, Ph.D., senior vice president.* Wayne is a specialist in strategic management, planning, product development, and quality. He directs planning, strategic management, and quality functions, reporting to Mel. Currently, Wayne is also a business development consultant and nonfiction (popular science, business development) author.

- *Tom Lewis, vice president.* Tom is an experienced entrepreneurial, business, and finance specialist who owns or co-owns several other businesses (one jointly with Mel). Tom directs finance and administrative operations and provides major contributions to strategic management.

- *Jack Robinson, director of development.* Jack is a proven entrepreneur, company owner, and product development, design, and production expert for fabric-based products. Jack oversees all in-house and contract fabric-related development, design, and production activities. He also provides significant contributions to strategic management.

- *Marc Ranger, director of engineering.* Marc is a senior electrical and electronics engineer with broad experience in modern electronic design. Marc's expertise is highly unusual in that it includes both analog and digital domains. Marc conducts or directs all in-house and contracted electronic development, design, and production and plays a major role in new-product conceptualization.

The ThermoGear™ board of directors currently includes Mel Campf, Wayne Fields, Tom Lewis, and Jack Robinson. A fifth board position has been reserved for a major strategic partner. A strong candidate is the president of the New York group with which ThermoGear™ has been negotiating long-term blanket production and marketing arrangement, or a representative of one (of two) backup blanket production and marketing entities.

The present ThermoGear™ board violates the dictum in this book of distinct management and board. At most, the president normally would serve both groups. ThermoGear™ is a special case, a reminder that exceptions work and can prove advantageous if they are not detrimental to company viability and performance. As a primary acceptability test, exceptions to the book's recipe are tentatively acceptable if funding impact is neutral or positive. The secondary acceptability test of an exception applies the same logic to the other six success factors.

ThermoGear™ has monitored the two exception-acceptability tests closely. The effects of the primary test have been favorable, since funding has been acquired at every requisite stage, in full and on time. All effects of the secondary test also have been positive or neutral. Management concludes that for ThermoGear™, excepting the board-management distinction dictum has conveyed favorable company impact.

Two comments are in order. First, ThermoGear™ continues to track the board-management distinction issue and stands ready to install a distinct board if changing conditions dictate. Second, the ThermoGear™ exception is not meant to encourage cavalier interpretation of the recipe put forth in this book. Rather, take the recipe seriously; thoroughly examine and monitor any rare exceptions for acceptability before and continuously after implementation.

B.4 Planning, Plan, and Funding

Planning, the business plan, and funding have been paramount for Thermo-Gear™, as expected. Planning and funding are identified success factors, and the business plan is a direct written expression of select planning elements.

The best of plans at one time can look much different with added progress and insight. No planning is perfect, and constant reexamination should be the rule. ThermoGear™ is a prime example. The first design effort, the bench prototype, was an incomplete design and effectively represented a terminated R&D effort. The design company prototype (second-generation design) also was incomplete but was eventually carried by in-house engineering (Marc) to a brief market test. Nevertheless, that design was rejected as well when national retailer feedback revealed requisite design specs the old design, even in modified form, could not meet. The real ChillBuster® design was the third try from scratch.

Some might mistakenly apply hindsight to say that ThermoGear™ could have avoided multiple designs. However, the future is not always so clear at each planning juncture, so analysis is instructive. The company's management team formed slowly. The second design generation (design company prototype) was half over before the team totaled two, and a team of three launched the third-generation design. The remaining management team members came on board after final design approach commitment (Marc Ranger had served as a consultant for some time before actually joining the team). Had ThermoGear™ had the benefit of all five team members in 1995, certainly the effort, cost, and product and business development time could have been lessened considerably. However, the real world works like evolution—you make do with what is available, not what might be later.

The planning and funding at ThermoGear™ has followed several excellent principles and featured winning choices at prominent decision points. We look first at key principles.

- *Team-sourced funding.* The first three team members had personal or direct access to funding sources covering all company needs to date.
- *Strategic partners.* The team sought strategic partners to address major resource needs (e.g., design, manufacturing, marketing, sales) rather than attempting to build all capabilities in-house from scratch.
- *Sweat equity.* To preserve cash, team members have built the product and the company with only minor expense reimbursement and no salaries.
- *Market-driven design.* ChillBuster® product design has depended almost totally on seller or end-user inputs to define the marketing and design specs.
- *Team concept.* Mel sensed from the start that success depended on team effort to take advantage of diverse expertise and skills, and the expanding team has retained that fundamental appreciation.

Those broad principles are generic and applicable to virtually any project. Following is a view of several ThermoGear™ choices of a much more project-dependent nature.

- *Bootstrap operation.* ThermoGear™ was and still is a garage operation, with all players operating from home or existing work facilities.

- *No salaries.* From the start, team members have drawn no salary, nor is any debt accumulating for time reimbursement later. All personal benefit derives from company common stock.

- *Contract design.* All design work was contracted, until in-house management team expertise was secured to carry out development and design.

- *Contract manufacturing.* Congruent with all formal planning, management has employed contract manufacturing and intends the same for foreseeable activity.

- *Marketing group.* Established marketing and sales channels were sought to provide rapid market access (everyone in the chain must receive suitable margins).

- *Molded control module.* Adopting the third-generation design for major market entry was a drastic change, but one that fit neatly with strategic marketing and sales partner alignment.

- *Quality commitment.* The company has committed to the business development recipe in this book, including the four control system complex described in Chapter 13.

- *Informed stockholders.* ThermoGear™ has carefully nurtured all stockholders and periodically kept them thoroughly abreast of every company development.

Absolute adherence to that list is not necessary, and additional considerations certainly will apply to many projects. Nevertheless, ThermoGear™ has employed multiple creative solutions for major hurdles that bear admiration and serious consideration.

B.5 Product Development and Design

Product drives every product-oriented growth company. Stadium blanket development reveals many real-world factors pertinent to understanding entrepreneurial product development. This section addresses details and rationale of several key ChillBuster® elements.

B.5.1 Product Concept

The ChillBuster® blanket inventor, Mel Campf, perceived a need and set out to fill it. He first searched for existing commercial products but found none. He then outlined a seemingly responsive product and examined subsystems for feasibility. It appeared on first blush that a functional product was possible.

From the moment a product was seriously considered, Mel envisioned a company to design, produce, and market any resultant device. Logically, however, he did not dwell on the company, because a product must come first. That was sound insight for anyone, especially a nontechnical person, a tribute to Mel's business and marketing acumen in two ways. First, he recognized that a product concept is far short of a real product. Second, he understood that a product needs a company to effectively organize its development and commercialization.

B.5.2   R&D

Mel next sought technical help. At the time, battery technology seemed key. Continued networking directed Mel to a gentleman experienced in battery-powered devices. That person, with Mel's guidance, led to the breadboard (first-generation) system, an electronic subsystems bench test to demonstrate system feasibility. However, it lacked the coherence of a full prototype system.

The first-generation prototype never saw daylight, but it carried significant impact. Certain technical solutions proved central to the current third-generation design and all pending patent actions. The latter design elements are proprietary and cannot be discussed until all patent actions have been completed. We also cannot know whether those design concepts would have arisen if ThermoGear™ in 1995 had proceeded directly to the full management team and third-generation design.

The first-generation prototype represented critical R&D, and the second-generation prototype also proved indispensable. The professional design company that produced the second-generation design and prototype added another patent-relevant technical solution. Furthermore, even though a prototype rather than a production system, the second-generation design proved sufficient for an identified contract manufacturer to make production ready. ThermoGear™ took that opportunity to test market an introductory product in late 1997, thereby gaining invaluable production and marketing experience.

B.5.3   Design, Manufacture, and Marketing

ThermoGear™ originally planned initial product to employ a fabricated aluminum control-module enclosure. Product revenues would be used in part to develop an injection-molded control module, leading to large-volume sales at a lower price. Injection molding requires expensive tooling. The hope was to demonstrate product and company viability and marketplace product acceptance with early expensive product, to attract redesign funding for subsequent high-volume production and marketing.

The strategic plan was sound, but management also pursued another parallel avenue. That created flexibility, in that secondary opportunities might allow accelerated company development. The idea was to seek strategic partners for joint stadium blanket commercialization. One or more strategic partners could benefit ThermoGear™ in the three realms of design, production, or marketing.

Several candidates presented in various ways, but especially attractive was a New York marketing group. ThermoGear™ and the New York group are actively detailing a long-term arrangement mutually beneficial and attractive to both.

ThermoGear™ is fully committed to any satisfactory New York arrangement. Concurrently, management wisely is pursuing other avenues, in case the New York arrangement is deficient or fails. One exciting new strategic partner candidate recently surfaced, and tertiary possibilities are being pursued as well, to better protect the company and its stockholders. The New York partner agrees that additional partners may be advantageous, so a multiple-way arrangement is possible.

The apparent superiority of potential strategic partnerships has led management to change course. Before, the plan was for in-house production and marketing, and strategic partners were backup. Now, strategic partners are the plan, and in-house production and marketing form the fallback. Either way, in-house ThermoGear™ designs drive the process. Flexible strategic posturing has allowed the company to pursue various mixes of in-house, contracted, and strategic partner solutions to manufacturing and marketing, without impeding design. Pliable strategic planning has postured ThermoGear™ to eventually pick the best of several pathways, while simultaneously cultivating several such opportunities. Also, the approach is low risk for the stockholders, compared to all in-house manufacturing and marketing, due to far lower resource requirements and availing existing operations.

The ChillBuster® blanket was originally conceived as a stadium (spectator event) blanket. Since then, many other blanket applications have been developed, details held proprietary until commercialization, as have ancillary products to various blanket versions. Those applications and ancillaries must remain proprietary for now. One involves a second ChillBuster® product that is largely designed. The point is that these factors supply critical leverage for an embryonic company, providing extended design applicability, growth potential, and company investor attractiveness.

B.5.4 Patents

Several unique design features have arisen through the three generations of ChillBuster® blanket design. Exact design features and patent strategy are proprietary, but generalities are instructive for general product protection.

There are two classes of patent protection, domestic and foreign. A single domestic patent commonly costs $5,000 to $15,000, plus ongoing maintenance fees. Many companies seek only a domestic patent, for economy. Foreign patents can exceed $100,000 for reasonably broad coverage.

Foreign patents often are prohibitively expensive, and such costs for initial products arise when cash is lean, prior to major product sales anywhere. However, the world's patent laws are paradoxical. A domestic patent application can be submitted up to a year after initial product promotion or sale. Strangely, foreign patent applications typically must be submitted prior to any promotion or sale.

One way around direct domestic or foreign patent costs is to recruit strategic partners willing to absorb the costs. Another technique for foreign patents is to use a Patent Cooperation Treaty (PCT) filing. For several thousand dollars, a PCT filing postpones detailed foreign patent filings for 18 months, allowing more time for revenue, investor, or strategic partner pursuit. In any case, due product protection attention must accompany early strategic planning.

Product protection attention also must precede any discussions with outside entities. It is best to file patent applications prior to outside discussions. Even a U.S. filing does not provide protection from someone else filing or manufacturing overseas. Nondisclosure agreements are less substantial, but they can be effective under sound legal advice. In any case, consult legal counsel and thoroughly understand relevant risks and benefits before undergoing any outside discussions.

B.5.5 Trademarks

There are fundamentally two types of trademark, common-law and registered. ThermoGear™ employs one of each type.

ThermoGear™ uses the TM superscript symbol by the company name. This represents a common-law trademark, assuming the symbol is universally displayed by the company. The advantage is instant use and no application submission process, but a common-law trademark is not registered with the U.S. Patent and Trademark Office. Use of this designation wherever the ThermoGear™ name appears places the world on notice that the company name is also viewed as a trademark. Thus, ThermoGear™ can be used to convey product or functional use meaning, in addition to the formal company name. Viewmaster exemplifies a historical company name that additionally conveyed product significance.

ThermoGear™ formally applied to the U.S. Patent and Trademark Office to register the ChillBuster® trademark and notice of successful registration has been received. The company intends to link the ChillBuster® designation long term to an entire product line. It is important to note that registration does not establish absolute trademark protection rights, which require actual consumer understanding. The test is whether the trade name (here, ChillBuster®) becomes specifically branded to distinct ThermoGear™ products by consumers. Registration is not essential but it is beneficial. Registration places the trademark claim on public record, grants benefits in legal proceedings, and allows identifying the mark as registered with the U.S. Patent and Trademark Office; proper commercial application can render the trademark incontestable.

B.6 Review of Success Factors

A final ThermoGear™ analysis is educational. The seven success factors introduced in Chapter 2 are universal to business development success. This section tests ThermoGear™ against each success factor standard.

B.6.1 Perspective

Management has exhibited outstanding perspective to date. Clear understanding of all short- and long-term relationships and consequences is apparent, through retention of broad internal consistency and coordination in spite of sweeping course changes. For instance, early planning had defined in-house design, manufacturing, and marketing as the company approach. However, strategic partners were seriously considered as an alternative, and when events favoring the latter approach arose, a normally drastic change became more a shift of attitude. The same flexible perspective relates to successive design generations, ongoing strategic planning, and funding, all functions that smoothly contended with dynamic, striking change.

ThermoGear™ management serves as a stunning example of what project leadership requires. True management, including the strategic sense, must exude deep and broad perspective to properly integrate the diverse challenges and opportunities projects face. There is only one way to acquire that global expertise and common sense—you must recruit it in full bloom. It is impossible for underdeveloped people to learn the requisite knowledge and experience on the fly.

B.6.2 Planning

ThermoGear™ has conducted continuous planning since its inception. There has been an incessant, pervasive professional dialog among the management team members at every step. Consensus is obtained for all planning detail, and no one makes changes without renewed team agreement.

Most important, the ThermoGear™ management team works well together. All players know each other's strengths and seek appropriate expertise as needed. Egos are left elsewhere. It often is hard to say who came up with an idea, because productive interchange fleshes out novel notions so quickly and thoroughly that original thoughts are obscured. Mutual respect is patently evident. These factors combine to create confident plans of action, because each player knows the group's collective strength. That is what teams are all about.

B.6.3 Management

ThermoGear™ management exhibits tremendous administrative depth, planning acumen, people skills, and leadership. Every action is considered in terms of global effect and immediate, short-term, and long-term consequences. Each management team member thoroughly knows the plan and conducts action consistent with plan advancement.

Build a team that functions with reason and grace, based on broad and deep knowledge and experience relevant to your industry. Fall short, and countless time, effort, and emotion will be spent reconciling clashes and addressing disagreements. Furthermore, no amount of effort will likely suffice, such problems being team inherent. Prevent this disastrous situation from ever arising. Follow Section 5.2 to the letter and build the management team correctly the first time.

B.6.4 Funding

Project funding is always overriding, ThermoGear™ included. The company has been successful, garnering investment well into six figures. Two points are pertinent. First, both macro- and microdetails of ThermoGear™ funding have been accomplished by following this book's recipe with professionalism and style. Second, ThermoGear™ spending needs may be about over, given consummation of one or more strategic partnership arrangements.

Securing the right money at the right time throughout business development is a monumental accomplishment worthy of strong praise. The many funding stages are so critical to business birth and growth that little proceeds otherwise. Credit for company success certainly goes to the outstanding ThermoGear™ management team and their ongoing attention to every detail.

This is a stunning example of powerful strategic planning and management in action.

B.6.5 Product

ThermoGear™ has a winning product set, on paper at least. The ChillBuster® blanket product line alone has enough distinct products and markets to sustain intermediate-term growth. The company also has additional ChillBuster® and other products in planning or development. Again, we see a commanding illustration of the recipe presented in this book.

The planned ThermoGear™ lineup features several desirable factors. For one, the line is simple, requiring nothing exotic in design, materials, or production. In addition, diversity is product inherent, requiring minor design tuning to address many markets. Also, the potential markets are large enough that the company can avail itself of volume production economies. Finally, alignments with strategic partners portend immediately tapping potent existing marketing and sales systems. ThermoGear™ is actually much stronger here than its superficial appearance, but additional details are proprietary. Nevertheless, plenty is visible to render ThermoGear™ a fine achievement standard.

B.6.6 Marketing

Marketing is the sole source of noninvestment money. The company plan to form strategic partnerships with established marketing channel entities provides fast and large sales potential. The alternative, building an in-house marketing program from scratch, is meaningfully more difficult and risky, at least for ThermoGear™. However, all products and markets are different. Creatively plan relative to your unique product and situation, as ThermoGear™ did. The real secret is putting together an outstanding management team, so you can successfully tackle anything realistic.

B.6.7 Location

Smoothly functioning ThermoGear™ management is particularly impressive, given the dispersed locations. In essence, ThermoGear™ is still a garage operation, everyone working part time without pay and using home or alternative business environments. Three team members are scattered around the greater area of Portland, Oregon. The other two team members live 300 miles away, in opposite directions, and make occasional Portland trips for requisite direct interactions.

The strategic partnership approach for manufacturing and marketing conveys ThermoGear™ advantage. Location is no longer a strategic consideration, the strategic partners assuming said requirements. Operating on its own, the company's West Coast location would be far from the domestic population center and quite remote from the large European market. Similar considerations may affect your product. Strategic partners or contract manufacturing or marketing can sometimes provide substantial strategic advantages.

B.7 Conclusion

We have examined ThermoGear™ in depth, because its situation clearly illustrates so many items of this book's recipe. It is advantageous to present diverse elements in the context of one project, to better show relationships.

The ThermoGear™ approach is not perfect. Still, ThermoGear™ is close to the minimum effort that will succeed, another reason for its review. You absolutely must acquire proper resources to achieve success. In turn, success requires meticulous attention to every success factor, especially to planning, the plan, funding, and management. Given powerful success factor adherence, almost everything else follows.

Besides the success factors, a lesson from ThermoGear™ is intentional strategic flexibility. One constant of business is that nothing is constant in business. ThermoGear™ intentionally positioned itself to follow any of several paths and then actively cultivated opportunities along the same paths. Management did not need to know what was best from the start. Creative strategic planning and management reduced the task to picking the best of contrasting opportunities, a much easier and less risky job. Of course, ThermoGear™ is still a work in progress, so the final result is not in. It does not matter here. The results to date suffice for ample examples of sound perspective, strategy, and execution.

Suggested Reading

Adbelmonem, A. A., *Computer-Aided Multivariate Analysis.* Chapman & Hall, 1996.

Arkebauer, J. B., *The McGraw-Hill Guide to Writing a High-Impact Business Plan: A Proven Blueprint for First-Time Entrepreneurs.* McGraw-Hill, 1994.

Baird, M., *Engineering Your Start-Up: A Guide for the Hi-Tech Entrepreneur.* Belmont, CA: Professional Publications, 1997.

Baty, G. B., *Entrepreneurship for the Nineties.* Prentice Hall, 1990.

Bell, C. G., *High-Tech Ventures: The Guide for Entrepreneurial Success.* Addison-Wesley, 1991.

Besant, C. B., and C. W. K. Lui, *Computer-Aided Design and Manufacture,* 3rd ed. Prentice Hall, 1986.

Biekert, R., *CIM Technology: Fundamentals and Applications.* Goodheart-Willcox, 1998.

Blohowiak, D., *Your People Are Your Product: How to Hire the Best So You Can Stay the Best.* Chandler House, 1998.

Bobrow, E. E., *The Complete Idiot's Guide to New Product Development.* Macmillan, 1997.

Boston Consulting Group (C. W. Stein and G. Stak, eds.), *Perspectives on Strategy: From the Boston Consulting Group.* Wiley, 1998.

Boyd, H. W., and O. C. Walker, *Marketing Management: A Strategic Approach.* Irwin, 1990.

Brandt, S. C., *The Ten Commandments for Building a Growth Company.* New American Library, 1983.

Breen, G., and A. B. Blankenship, *Do-It-Yourself Marketing Research.* McGraw-Hill, 1989.

Carey, G. F., *Circuit, Device, and Process Simulation: Mathematical and Numerical Aspects.* Wiley, 1996.

Capezio, P., and D. Morehouse, 1993. *Taking the Mystery out of TQM: A Practical Guide to Total Quality Management.* Hawthorne, NJ: Career Press, 1993.

CGMP (Current Good Manufacturing Practices), *Medical Device Quality Systems Manual: A Small Entity Compliance Guide,* 1st ed. Updated Jan. 14, 1997. Download from Internet URL http://www.fda.gov/cdrh/gmp_man.html.

Chang, T-C., R. A. Wysk, and H-P. Wang, *Computer-Aided Manufacturing*. Prentice Hall, 1997.

Chapman, R. G., *Brandmaps: The Competitive Marketing Strategy Game*, 4th ed. Prentice Hall, 1996.

Chisnal, P. M., *Strategic Industrial Marketing*. Prentice Hall, 1985.

Chisnal, P. M., *The Essence of Marketing Research*. Prentice Hall, 1991.

Cohen, W. A., *Model Business Plans for Product Businesses*. Wiley, 1995.

Cooper, R. G., *Product Leadership: Creating and Launching Superior New Products*. Perseus, 1998.

Cooper, R. G., E. J. Kleinschmidt, and S. J. Edgett, *Portfolio Management for New Products*. Addison Wesley Longman, 1998.

Crosby, P., *Let's Talk Quality*. McGraw-Hill, 1989.

Davidow, W. H., *Marketing High Technology*. Free Press, 1986.

Davis, W. S., *Business Systems Analysis and Design*. Course Technology, 1994.

Deming, W. E., *Quality, Productivity, and Competitive Position*. MIT Press, 1982.

Dibb, S., L. Simkin, W. M. Pride, and O. C. Ferrell. *Marketing: Concepts and Strategies*. Houghton Mifflin, 1991.

Dobson, P., and K. Starkey, *The Strategic Management Blueprint*. Blackwell, 1993.

Drucker, P. F., *Innovation and Entrepreneurship*. Harper & Row, 1985.

Ennew, C. T., *The Marketing Blueprint*. Blackwell, 1993.

Field, D., *Take Your Company Public—the Entrepreneur's Guide to Alternative Capital Resources*. Simon & Schuster, 1991.

Fields, R. W., *God, Cosmos and Man: The Role of Mind in a Purposeful Universe*. Medicine Bear, 1998.

Ford, D. (ed.), *Understanding Business Markets*. Academic Press, 1990.

Fox, J. J., *How to Become CEO: The Rules for Rising to the Top of Any Organization*. Hyperion, 1998.

Gajski, D. D., *Principles of Digital Design*. Prentice Hall, 1996.

Galloway, R. L., *Statistics for Marketing and Business*. Stanley Thornes, 1989.

Galloway, R. L., *Principles of Operations Management*. Routledge, 1993.

Garrett, T. M., and R. J. Klonoski, 1986. *Business Ethics,* 2nd ed. Prentice Hall, 1986.

Gill, J. O., *Understanding Financial Statements: A Guide for Non-Financial Readers*. Crisp Publications, 1990.

Goodstein, L. D., J. W. Pfeiffer, and T. Nolan, *Applied Strategic Planning: A Comprehensive Guide*. McGraw-Hill, 1993.

Gorchels, L., *The Product Manager's Handbook*. NTC Publishing Group, 1995.

Gregory, J. R., and J. G. Wiechmann, *Marketing Corporate Image: The Company as Your Number One Product*, 2nd ed. NTC Publishing Group, 1998.

Grove, A. S., *High Output Management*. Random House, 1983.

Hamper, R. J., and L. S. Baugh, 1990. *Strategic Market Planning*. NTC Business Books, 1990.

Harvard Business School, *Harvard Business Review on Managing Uncertainty*. Harvard Business School, 1999.

Hawken, P., *Growing a Business.* Simon & Schuster, 1987.

Hill, T., *Manufacturing Strategy.* Irwin, 1989.

Hill, T., *Production Operations Management.* Prentice Hall, 1991.

Hirshberg, J., *The Creative Priority: Driving Innovative Business in the Real World.* Harper, 1998.

Hollins, G., and B. Hollins, *Total Design.* Pitman, 1991.

Holtzschue, L., *Design Fundamentals for the Digital Age.* Wiley, 1997.

Horn, T., *Business Valuation Manual: An Understandable, Step-By-Step Guide to Finding the Value of a Business—Revised.* Charter Oak Press, 1987.

Juran, J. M., *Quality Control Handbook,* 4th ed. McGraw-Hill, 1988.

Juran, J. M., and F. M. Gryna, *Quality Planning and Analysis: From Product Development through Use,* 3rd ed. McGraw-Hill, 1991.

Kalb, I. S., *Selling High-Tech Products and Services.* K&A Press, 1990.

Kaplan, R. S., and D. P. Norton, *The Balanced Scorecard: Translating Strategy into Action.* Harvard Business School, 1996.

Katsaros, J., *Selling High Tech/High Ticket: Using Relationship Management Techniques to Sell and Service Today's Complex Products.* McGraw-Hill, 1993.

Katzenbach, J. R., and D. K. Smith, *The Wisdom of Teams: Creating the High-Performance Organization.* Harper, 1994.

Kawasaki, G., and M. Moreno, *Rules for Revolutionaries: The Capitalist Manifesto for Creating and Marketing New Products and Services.* Harper, 1999.

Kelle, U., G. Prein, and K. Bird, *Computer-Aided Qualitative Data Analysis: Theory, Methods and Practice.* Sage Publ., 1995.

Kotler, P., *Marketing Management: Analysis, Planning, Implementation and Control,* 7th ed. Prentice Hall, 1991.

Kotter, J. P., *Leading Change.* Harvard Business School, 1996.

J. K. Lasser Institute, *How to Run a Small Business,* 4th ed. McGraw-Hill, 1974.

J. K. Lasser Institute, *How to Read a Financial Statement.* Simon & Schuster, 1984.

Lewis, C. D. *Industrial and Business Forecasting Methods.* Butterworths, 1982.

Lewis, C., *Demand Forecasting and Inventory Control: A Computer-Aided Learning Approach.* Wiley, 1998.

Lewis, J. W., *Modeling Engineering Systems: PC-Based Techniques and Design Tools.* LLH Tech Publ., 1993.

Lipper, A., III, and G. Ryan, *Venture's Guide to Investing in Private Companies: A Financing Manual for the Entrepreneurial Investor.* Irwin, 1984.

Lizardy, A., *The 10 Cornerstones of Selling: How to Get Greater Control of Your Selling Results.* Avant Books, 1992.

Lockyer, K. G., *Critical Path Analysis and Other Project Network Techniques.* Pitman, 1984.

Lockyer, K. G., *Production and Operations Management.* Pitman, 1990.

Lorenz, C., *The Design Dimension.* Blackwell, 1986.

MacDonald, M., *Marketing Plans: How to Prepare Them, How to Use Them.* Heinemann, 1989.

McGrath, K., and S. Elias, *Trademark: Legal Care for Your Business and Product Name.* Nolo Press, 1997.

McGrath, M. E., *Product Strategy for High-Technology Companies: How to Achieve Growth, Competitive Advantage, and Increased Profits.* McGraw-Hill, 1994.

McVicker, M. F., *Small Business Matters: Topics, Procedures, and Strategies.* Chilton, 1990.

Meyer, M. H., and A. P. Lehnerd, *The Power of Product Platforms: Building Value and Cost Leadership.* Free Press, 1997.

Monger, R. F., *Mastering Technology.* Free Press, 1988.

Moore, G. A., *Crossing the Chasm: Marketing and Selling High-Tech Products to Mainstream Customers.* Harper, 1995.

Morgan, A., *Eating the Big Fish: How Challenger Brands Can Compete Against Brand Leaders.* Wiley, 1998.

Nanus, B., *Visionary Leadership: Creating a Compelling Sense of Direction for Your Organization.* Jossey-Bass, 1992.

Nesheim, J., *High Tech Startup: The Complete How-To Handbook for Creating Successful New High Tech Companies.* Electronic Trend Publications, 1992.

Norman, R., *Service Management.* Wiley, 1991.

O'Donnell, M., *Writing Business Plans That Get Results.* Contemporary Books, 1991.

Ostroff, F., *The Horizontal Organization: What the Organization of the Future Looks Like and How it Delivers Value to Customers.* Oxford University Press, 1999.

Owen, R. R., D. R. Garner, and D. S. Bunder, *The Arthur Young Guide to Financing for Growth.* Wiley, 1986.

Peach, R. W. (ed.), *The ISO 9000 Handbook.* McGraw-Hill, 1996.

Pettis, C., *TechnoBrands: How to Create and Use "Brand Identity" to Market, Advertise and Sell Technology Products.* AMACOM, 1994.

Pierce, J. A., II, and R. B. Robinson, Jr., *Formulation and Implementation of Competitive Strategy,* 2nd ed. Irwin, 1985.

Porter, M., *Competitive Strategy: Techniques for Analyzing Industries and Competitors.* Free Press, 1980.

Porter, M., *Competitive Advantage: Creating and Sustaining Superior Performance.* Free Press, 1985.

Rasiel, E., *The McKinsey Way: Using the Techniques of the World's Top Strategic Consultants to Help You and Your Business.* McGraw-Hill, 1999.

Reinertsen, D. G., *Managing the Design Factory: The Product Developer's Toolkit.* Free Press, 1997.

Rich, S. R., and D. E. Gumpert, *Business Plans That Win $$$: Lessons From the MIT Enterprise Forum.* Harper & Row, 1987.

Ries, A., and L. Ries, *The 22 Immutable Laws of Branding: How to Build a Product or Service into a World-Class Brand.* Harper, 1998.

Roberts, E. B., *Entrepreneurs in High Technology: Lessons From MIT and Beyond.* Oxford University Press, 1991.

Ronstadt, R., and J. Shuman, *Venture Feasibility Planning Guide: Your First Step Before Writing a Business Plan.* Dana Point, CA: Lord Publishing, 1988.

Schnaars, S., *Marketing Strategy: A Customer Driven Approach.* Free Press, 1991.

Schonberger, R. J., *World Class Manufacturing: The Next Decade: Building Power, Strength, and Value.* Free Press, 1996.

Sletten, E., *How to Succeed in Exporting and Doing Business Internationally.* Wiley, 1994.

Slywotzky, A. J., B. Andelman, and D. J. Morrison, *The Profit Zone: How Strategic Business Design Will Lead You to Tomorrow's Profits.* Times Book, 1997.

Smith, R. F., *Entrepreneur's Marketing Guide: Bigger Sales Power for Smaller Firms.* Reston Publishing, 1984.

Sobel, M., *The 12-Hour MBA Program.* Prentice Hall, 1994.

Stewart, J. D., *The Power of People Skills.* Wiley, 1986.

Strock, J. J., *Engineering for Profit: Successful Marketing of Hi-Tech Products and Systems.* CRC Press, 1994.

Suzaki, K., *The New Manufacturing Challenge: Techniques for Continuous Improvement.* Free Press, 1987.

Tarrant, J., *Perks and Parachutes.* Linden Press/Simon & Schuster, 1985.

Thielen, S., and D. Thielen, *The 12 Single Secrets of Microsoft Management: How to Think and Act Like a Microsoft Manager and Take Your Company to the Top.* McGraw-Hill, 1999.

Thomas R. J., *New Product Development: Managing and Forecasting for Strategic Success.* Wiley, 1993.

Tibbetts, J. S., Jr, and E. T. Donovan, "Compensation and Benefits for Startup Companies: How to Conserve Cash and Still Attract the Best Executives." *Harvard Business Review,* Jan.–Feb. 1989: 140–47.

Tooley, D. F., *Production Control Systems and Records.* Gower, 1985.

Ulrich, K. T., and S. D. Eppinger, *Product Design and Development.* McGraw-Hill, 1994.

Urban, G. L., and J. R. Hauser, *Design and Marketing of New Products,* 2nd ed. Prentice Hall, 1994.

Usher, J. M., H. R. Parsaei, and U. Roy (eds.), *Integrated Product and Process Development: Methods, Tools, and Techniques.* Wiley, 1998.

Vesper, K., *New Venture Strategies,* rev. ed. Prentice Hall, 1990.

Wheeler, D. J., and D. S. Chambers, *Understanding Statistical Process Control.* SPC Press, 1998.

White, R. M., *The Entrepreneur's Manual.* Chilton, 1977.

Whiteley, R., and D. Hessan, *Customer Centered Growth: Five Proven Strategies for Building Competitive Advantage.* Addison Wesley Longman, 1997.

Wiener, D., and N. Glaskowsky, *Theory and Problems of Business Law.* McGraw-Hill, 1985.

Wild, R., *Operations Management—A Policy Framework.* Pergamon, 1980.

About the Author

Dr. R. Wayne Fields spent the better part of college (Oregon State University, Corvallis) training as a physics major. In 1963 he finished a bachelor's degree in science by building a companion background in biological disciplines. Upon graduation, his combination of physical and biological sciences led to a graduate studies scholarship at the University of Oregon Medical School (now the Oregon Health Sciences University), where in 1969 he obtained a Ph.D. in physiology.

Trained in a fashion that normally serves as a prelude to teaching, Dr. Fields instead embarked on a career in research. His interests centered on applied research and investigations on topics with promise of functional implementation rather than studies directed purely to extend knowledge (basic research). His interest in practical applications continued and after many years led to a major career shift—an R&D management role in the medical device industry. R&D sparked entrepreneurial interests, and many years later Dr. Fields undertook an executive role in an entrepreneurial business for a time. Then, in the late 1980s, he helped found and run an unusual business development firm. That entity, Venture Solutions, Ltd. (VSL) of Lake Oswego, Oregon, still exists, even though the original co-owners have largely moved on to other associated activities. (VSL's telephone number is 800-346-2408.)

VSL is crucial to the evolution of this book. Dr. Fields and the VSL co-owner, Norm Locati, both have technical backgrounds (Mr. Locati is an industrial and mechanical engineer) and both brought to VSL substantial R&D and product-oriented entrepreneurial and executive management experience. While prior activities provided important knowledge and experience, VSL formed the real-world nucleus from which the business development theory and practice of this book arose. In a sense, VSL has provided a venture capital source for the little guy, addressing projects with perhaps substantial potential but lacking the prowess and appeal of the traditional venture capital market. The approach was unique in three ways: (1) each project was shopped independently rather than being funded from a central capital pool; (2) subject projects typically were too small for traditional venture capital interest; and (3) the VSL team often played an active role in project grooming and development. That hands-on, deep involvement in individual projects residing below the well-publicized, glamorous venture capital tier led to a unique perspective on business development and the distinctive approach revealed in *The Entrepreneurial Engineer.*

The author currently has retired from the daily grind of the project stream to allow personal involvement in select projects.

Index

Acceptance activities, 187
Accounting
  accrual, 82
  activities, 76
  cash, 82
  professionals, 255
Accrual accounting, 82
Actual cost method
  defined, 103
  percentage cost method vs., 104
Administration, 31, 45
  activities, 233–34
  competitive position and, 220
  design and, 174–75
  during R&D, 151–52
  employee development/relations and, 223
  issues, 142
  product introduction and, 194, 196–97
  productivity and, 219
  profitability and, 218
  prototyping and, 169
  public responsibility and, 223
  strategic issues in, 206–8
  technological leadership and, 222
  See also Marketing; Technology division
Administrative feasibility, 55–56
Administrative preparation, 182–86
  finance, 184–85

general management, 183–84
planning, 185
quality assurance, 185–86
See also Launch preparation
AIDA model, 228
Analytical feasibility tests, 146
Appearance model, 168
Appendixes, 96, 104–6, 290–94
  administration, 293
  content of, 105
  core, 291–93
  defined, 40, 96, 290
  general, 293–94
  historical financial data, 292
  literature, 294
  marketing, 293
  organization, 292–93
  photographs, 294
  primary financial projections, 291
  quality, 294
  resumes, 293
  sales, 293
  spectrum of, 105–6
  supportive financial projections, 291–92
  technical, 293
  See also Business plan
Appraisal costs, 249–50
Approvals, 167

319

Recent Titles in the Artech House Technology Management and Professional Development Library

Bruce Elbert, Series Editor

For further information on these and other Artech House titles, including previously considered out-of-print books now available through our In-Print-Forever® (IPF®) program, contact:

Artech House
685 Canton Street
Norwood, MA 02062
Phone: 781-769-9750
Fax: 781-769-6334
e-mail: artech@artechhouse.com

Artech House
46 Gillingham Street
London SW1V 1AH UK
Phone: +44 (0)20 7596-8750
Fax: +44 (0)20 7630-0166
e-mail: artech-uk@artechhouse.com

Find us on the World Wide Web at:
www.artechhouse.com